Essential Skills

for Science & Technology

D0771580

Essential Skills

for Science & Technology

Peter Zeegers • Kate Deller-Evans
Sandra Egege • Christopher Klinger

OXFORD
UNIVERSITY PRESS

253 Normanby Road, South Melbourne, Victoria 3205, Australia

Oxford University Press is a department of the University of Oxford.
It furthers the University's objective of excellence in research,
scholarship, and education by publishing worldwide in

Oxford New York

Auckland Cape Town Dar es Salaam Hong Kong Karachi
Kuala Lumpur Madrid Melbourne Mexico City Nairobi
New Delhi Shanghai Taipei Toronto

With offices in

Argentina Austria Brazil Chile Czech Republic France Greece
Guatemala Hungary Italy Japan Poland Portugal Singapore
South Korea Switzerland Thailand Turkey Ukraine Vietnam

OXFORD is a trademark of Oxford University Press
in the UK and in certain other countries

National Library of Australia Cataloguing-in-Publication data

Essential skills for science and technology.

Includes index.
ISBN 9780195558319 (pbk.).

1. Study skills. 2. Science—Study and teaching (Higher).
3. Technology—Study and teaching (Higher). I. Zeegers,
Peter.

600.0711

Edited by Lindsay Taaffe
Typeset by diacriTech
Proofread by Helen Yeates
Indexed by Russell Brooks
Printed in Hong Kong by Sheck Wah Tong Printing Press Ltd.

Brief Contents

Expanded Contents

List of Figures

Guided Tour

Key Concepts

Each chapter begins with an overview of the key points to aid understanding and navigation.

1

The First Step

Only fools are certain ... it takes wisdom to be confused.
Julius (Groucho) Marx, American comic (1890–1977)

Key Concepts

- Graduate attributes
- Transition to university
- Potential study problems
- Skills for academic success

Introduction

Congratulations on entering the world of higher education. As an individual, you have your own expectations and goals in life, and study may be an important component of these. What do you hope to gain from tertiary study? Is it a means to a future career? Are you here to broaden your mind? Are you here to gain an academic qualification? Or are you here until you decide upon your direction in life? Whatever your reasons for choosing to undertake higher education, you do so knowing that there are financial costs, such as fees, and that it requires a considerable commitment involving time and hard work.

Having chosen to study a course with a science component, a number of pathways are available to you. You may wish to work in fields as diverse as robotics, marine biology, computer animation, or the pharmaceutical industry. You may wish to obtain a professional qualification and embark on a career as a doctor, nurse, science teacher, speech pathologist, or electrical engineer. It may also be that a background in the sciences, with its requisite analytical and critical thinking skills, is the ideal background for a career in

Example

Scientific Officer (Nanobiotics)
Professional scale PSO1: $51 500–$59 500 pa
Ref. No: KL 103–07
KLATU Technology Corporation
Microrobotics division
Melbourne office

You look up the website for the KLATU Corporation and download the essential selection criteria to be addressed in the application, which may include:

- an appropriate degree in science or technology
- demonstrated skills in writing technical reports, grant applications, and publications
- the ability to work as a member of a team
- demonstrated ability to analyse complex research data
- the ability to develop and apply new research techniques to solve existing problems
- interpersonal and communication skills
- the ability to supervise technical staff.

So when the people at the KLATU Corporation look through the job applications the one thing that all applicants will have is an appropriate university qualification. What will set the applicants apart will be the ability of each to demonstrate their skills and abilities under each of the other selection criteria. Thus you can see the importance of all the things you learn at university.

This is just one example of the importance of graduate attributes, but the principle applies to most, if not all, professional positions, whether you are applying for a position as a physiotherapist, a teacher, a vet, or a mining engineer. Thus the goal of this book is to provide you with the means to develop essential academic and professional skills for use in science-based professions.

Transition to university

Going to university for the first time is much like starting a new job or moving to a new town. There is much that may be new to you at university, including the culture and the language of higher education. The types of problems encountered by students in any science-based course can be unique to that course, and to each student. However, many of the difficulties you face as a new student at university are common to most students, irrespective of whether you have just completed secondary school or have not studied for some time. Some of these are directly related to aspects of teaching and

Examples

Examples provide further explanation and demonstration of the key ideas.

20   Part 1: You and Tertiary Study

Procrastination

For the list of self-imposed factors above, perhaps the overriding problem is that of procrastination. Procrastination means delaying or avoiding something that must be done, and some of us develop this to an art form. The key to overcoming procrastination is to be decisive. Make decisions that will lead to positive outcomes rather than wasting time and energy thinking about them. The following strategies will help overcome a tendency to procrastinate:

1   Recognise that procrastination is a habit that cannot be changed unless *you* want to change it.

2   Be motivated to change. Self-belief is required to change habits.

3   Start small. Large goals are often difficult to reach and require concerted effort and time. Small, achievable goals are the best way to change your habits.

4   Keep going once small goals are reached. Setting and reaching goals becomes an ongoing process to success.

5   Failure is a part of success. Learn to accept that you cannot always achieve the ideal.

6   Reflect on the reasons for failure and use those insights to improve your next goal-setting session.

TEN TIPS ON BETTER TIME MANAGEMENT

There are many strategies used by successful students. The following is a list of some that may help you to be a successful time manager. Keep in mind that the key factor in all forms of management is attitude.

1   Set clear objectives and priorities and recognise the difference. For example, your objective may be to obtain a tertiary qualification to be a vet, but your priority is to maintain your health and fitness in the process.

2   In order to achieve your objectives you must decide which are the most effective activities that will help you achieve those goals. Remember that some tasks are a means to an end and some are an end in themselves.

3   Plan your time and your study. Have a long-term plan, a short-term plan, and a daily 'to-do list' in order of priority.

4   Stay positive. You can do this by making sure that during each day you accomplish something worthwhile and reward yourself for it.

5   Accept that you will make mistakes, learn from them, and don't repeat them. In planning your activities allow enough flexibility to accommodate mistakes, mishaps, and the unexpected.

6   Use all information sources so that you have the best possible chance of doing most things right the first time. In that way you won't have to waste time repeating things.

Tips

Tips provide useful advice throughout each chapter.

Boxes

Boxes highlight useful additional details about the topics to aid comprehensive understanding.

6   Part 1: You and Tertiary Study

Students' experiences of university

I came with much anxiety to Australia to study, but this week (Orientation Week) has helped put some of my fears at ease. The system here requires very much more of students, where back home the teachers tell us exactly what to study and how to study it. I now know that I will need to learn in a new way to become more independent just like the local students if I am to be successful.

Rangit, first-year Chemistry

I am finding it really hard going at the moment. I like having someone around to tell me whether I'm doing the right thing or if I'm on the right track. I find this really helps me. Also, as a student doing three lab courses I seem to spend a lot of time doing things without knowing why I am doing them nor understanding what it is that I am actually doing.

Celia, first-year Biology

University study is very hard. It is very demanding and time consuming if you do all that is required of you. I think a good student needs to learn self discipline and organisation. There are so many things to get used to. I guess you have to rely on yourself. It's really like a process of self discovery isn't it? It would be very helpful though to get more guidance and clear explanations of course requirements.

Carla, third-year Biology

I like university so far, it's better than school, really big though compared to the school I went to. One of the things I miss compared to school is that we don't really have any proper discussion times. We just seem to be always busy doing things. It would be nice to have more small-group tutorials so we can actually ask questions and not just answer problems. It's all just lectures and labs and computer stuff but little opportunity to actually ask questions.

Nick, second-year Information Technology

One of the key things I have learned over the last three years is to not give up when things get difficult. You need to learn to make sacrifices for the sake of your study and to keep your mind focused on why you are here. Go to all your lectures and tutorials and get as much information from other students as possible. You have to learn to accept failure as part of success. And also remember to have a good time.

Beverley, third-year Nursing

It's all about dedication and working your butt off. The most important thing I have learned is to take responsibility for myself and work things out. You have to. The faculty staff have their own things to do so they cannot look after you. Once you realise it's all up to you it's not too bad.

Paul, third-year Engineering

If we were to look at the common areas of difficulty facing new students we could group them accordingly:

· academic preparedness
· academic progress
· financial affairs

Summaries

A concise summary at the end of each chapter helps students to identify the most important points covered in the chapter.

---

7   Finish what you start. Some people (such as Leonardo da Vinci) jump from one activity to another until their objectives have been met. Others (such as Marcel Proust) tend to focus for extended periods on a single activity until it is complete, and only then will they move on to the next activity. It does not matter what approach you use as long as it works for you and you get things done.

8   Conquer procrastination. Be honest with yourself and eliminate or minimise time-wasting activities.

9   Accept that you cannot avoid the unexpected, so learn to handle things when they go wrong.

10 Enjoy university; it is a rewarding experience.

Summary

Studying requires good organisational and management strategies; you will be effective by:
*   being less rushed and have more time
*   keeping motivated to succeed
*   minimising the opportunities for procrastination and avoidance
*   avoiding meeting deadlines at the last minute
*   reducing your stress and anxiety levels to perform at your best.

A quote from Stephen Covey's book *Seven Habits of Highly Effective People* may put things in perspective:

The successful person has the habit of doing the things failures don't like to do. They don't like doing them either ... but their dislike is subordinate to the strength of their purpose.

Preface

For the great things of life are not done by a single impulse, but by a series of small steps brought together.

Vincent Van Gogh, letter to his brother Theo, 1889

This book is designed to assist you in the development of the academic skills required for the study of any science-based topic or course at the tertiary level. We are deliberately using the broadest definition of science to include the medical and human sciences (such as medicine, nursing, or psychology); the physical sciences (such as chemistry, geology, or physics); the biological sciences (such as ecology, marine biology, or zoology); and the technology areas (such as engineering, information technology, or nanotechnology). We have intended the book to be useful for students at all levels of tertiary study, but it is particularly suited to students in their first year at university.

Science is not simply a collection of discrete areas of study such as those mentioned above, it is an approach taken by those who strive for a greater understanding of the physical universe in which we live and our desire to know more about ourselves. Unfortunately today, science often tends to dominate the headlines when there is bad news, such as:

- Nuclear disaster at Chernobyl
- Mobile phones linked to cancer
- The Antarctic ozone hole growing larger.

Yet science surrounds us in our daily lives. There is science behind the methods we use to provide the best possible healthcare for all of our citizens. Science informs us of the nature of our own planet and provides us with the means to explore our celestial neighbours. Science provides us with the tools we use to evaluate the impact we have on our natural environment and our relationship to other living organisms. Science provides us with the things we use in our every day lives, including materials for clothing, shelter, and communications, to the drugs we take when we are ill. But above all, science is an activity undertaken by people, and in order for you to be able to participate fully in this activity you need a level of competence in a range of academic and professional skills.

The development of these key skills is the core goal of this book. The book content is divided into five sections: You and tertiary study; Teaching and learning; Research and critical appraisal; Writing and presenting; and Quantitative methods. Each chapter provides an overview of the topic under discussion through the development of a number of specific skills. If you find that you require more information, we have included a comprehensive bibliography at the end of the book, as well as a list of informative Websites.

The key to success at university is to be an effective independent learner, which involves taking charge of your own learning progress. Being an independent learner is a matter of self-reliance and confidence in your own abilities and strategies. One aspect of this, though it may appear to be contradictory to the concept of independence, is to know when, where, and how to go about seeking help when the need arises. In addition to the development of the skills mentioned above, studying also requires diligence, resourcefulness, and above all, perseverance. Remember that education is not just an end product, but rather, it is a process of discovery, development, and self-awareness.

Finally, this book is not a crime novel that you start reading on page one and find out 'who done it' on the last page. It is a book of resources and ideas based on the experience of many years of teaching in higher education by the authors, and on the collective experiences of many successful tertiary students. For this book to be effective we hope that you will refer to the different chapters as you need them and that in doing so it will stimulate you to know more. In the not too distant future, perhaps, your success at university may contribute to the success of future generations of students through subsequent editions of this book.

About the Authors

Dr Peter Zeegers, *PhD, GradDipEd, BSc(Hons), CertMedLabTech*. Peter Zeegers worked in medical research at the University of Adelaide and Flinders Medical Centre for more than ten years. He then moved to Flinders University where he has held positions as a lecturer in chemistry, academic adviser for the sciences, and Head of the Student Learning Centre. His research focuses on teaching and learning in higher education, and in particular the use of psychometric instruments to quantify the factors impacting on student learning, the development of causal models of learning outcomes, the relationship between psychological distress (depression, anxiety, and stress) and learning, and transition to university. He is the recipient of a National Teaching and Learning grant, several overseas fellowships, and is the author of two books and more than twenty research articles.

Ms Kate Deller-Evans, *BA(Hons), GradDipEd, AdvDipArts (ProfWriting)*. Kate Deller-Evans is the coordinator of Professional Writing at the Adelaide Centre for the Arts (TAFE). Prior to this appointment Kate was a lecturer in academic writing at Flinders University and the University of Canberra. Kate is a nationally recognised writer and poet, and has published widely with two collections of poetry, a creative writing manual, and eight books as editor. A member of the Australian Society of Editors, Kate is responsible for introducing editing to Flinders University's Research Higher Degree Professional Development Program. Kate's focus is in helping and encouraging all those who want to study or write. Kate is currently completing her doctoral studies at Flinders University and has begun writing the first part of a fantasy trilogy.

Ms Sandra Egege, *MA, BA (Hons), DipT, TESOL*. Sandra Egege is a lecturer in the Student Learning Centre at Flinders University. Sandra coordinates the academic stream of the Research Higher Degree Professional Development Program and lectures on research, academic writing, and critical thinking. During 1995–2001 she was the research and administrative officer for the Centre for Applied Philosophy. Sandra's research field is Philosophy of Mind, with an interest in the Philosophy of Science, Epistemology, and Philosophy of Biology. She has taught a broad range of topics in philosophy, including critical thinking, and formal and informal logic, and is a joint author of several published papers. She is working on a study guide on critical thinking and argumentation, and researching the efficacy of overtly teaching critical thinking to tertiary students.

Dr Chris Klinger, *PhD, BSc(Hons), GradCertTertEd, DipAdvSoundEng*. Chris Klinger is the Director of Foundation Studies at the University of South Australia and is a physicist with a PhD in theoretical cosmology. Prior to his current appointment he was a lecturer in the Student Learning Centre at Flinders University, as well as coordinator of the Flinders

Foundation Course, and a lecturer in the School of Commerce. Chris is particularly interested in helping students to overcome maths anxiety, to recognise and utilise the language aspect of mathematics, and to see mathematics and numeracy in a wider and more relevant context, thus promoting understanding, confidence, and more effective learning. His current research interests (outside of physics) are focused on examining the attitudes of adult learners towards mathematics and their mathematics self-efficacy beliefs.

PART 1

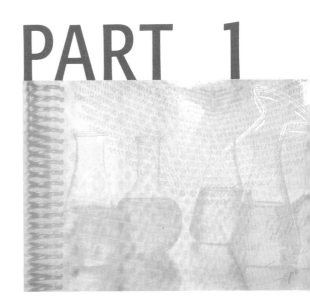

You and Tertiary Study

1

The First Step

Only fools are certain ... it takes wisdom to be confused.

Julius (Groucho) Marx, American comic (1890–1977)

Key Concepts

- Graduate attributes
- Transition to university
- Potential study problems
- Skills for academic success

Introduction

Congratulations on entering the world of higher education. As an individual, you have your own expectations and goals in life, and study may be an important component of these. What do you hope to gain from tertiary study? Is it a means to a future career? Are you here to broaden your mind? Are you here to gain an academic qualification? Or are you here until you decide upon your direction in life? Whatever your reasons for choosing to undertake higher education, you do so knowing that there are financial costs, such as fees, and that it requires a considerable commitment involving time and hard work.

Having chosen to study a course with a science component, a number of pathways are available to you. You may wish to work in fields as diverse as robotics, marine biology, computer animation, or the pharmaceutical industry. You may wish to obtain a professional qualification and embark on a career as a doctor, nurse, science teacher, speech pathologist, or electrical engineer. It may also be that a background in the sciences, with its requisite analytical and critical thinking skills, is the ideal background for a career in areas not usually associated with science, such as management, finance, or international relations. It is also quite common these days to study in a double degree program, for example science and law. Whatever area of study you choose, or your ultimate career goals, a university degree should not be seen as an end point but as a pathway to other career options and life experiences. In order to make the most of the first step into higher education and to make your time at university as productive as possible, you need to enhance and develop the essential skills for successful study.

There are many resources available to assist in your tertiary studies. In addition to the course-specific textbooks for your various topics, there is a wide variety of books and electronic sources, such as websites, to assist you to develop your learning skills. Some of these resources are listed in the appendices at the end of this book.

Graduate attributes

A university education in any area of the sciences is about the development of two types of skills and knowledge. First, the *specific* skills and knowledge relevant to your chosen field of study, and second, the range of *general* skills and knowledge expected of all university graduates, but which are also context dependent. These broader skills of university graduates are usually referred to as *graduate attributes*.

Examples of graduate attributes

- Literacy and numeracy
- Critical and analytical thinking, including problem solving
- Information literacy and research skills
- Written and oral communication within a professional context
- Management of time, resources, and tasks
- Cooperative and leadership skills as a member of a group

Graduate attributes are the skills you will develop as part of the university experience, and, as such, they are generally not explicitly taught as part of the set curriculum for your course. In his discussion of learning in the digital age, Alvin Toffler predicted in his book *The Third Wave* that 'the illiterates of the 21st century will not be those who cannot read and write, but those who cannot learn, unlearn, and relearn'. With this in mind, it would appear that in the future 'uneducated' will mean not knowing how to keep on learning, so, at university, great emphasis is placed on strategies for lifelong learning.

Imagine that you are a student studying nanotechnology and that there are a hundred students in your class. At the end of the final year of your course, you and your classmates, as well as students from other universities, will be looking for employment. Late in December there is an advertisement in *The Australian* newspaper for:

Example

Scientific Officer (Nanobiotics)
Professional scale PSO1: $51 500–$59 500 pa
Ref. No: KL 103–07
KLATU Technology Corporation
Microrobotics division
Melbourne office

You look up the website for the KLATU Corporation and download the essential selection criteria to be addressed in the application, which may include:

- an appropriate degree in science or technology
- demonstrated skills in writing technical reports, grant applications, and publications
- the ability to work as a member of a team
- demonstrated ability to analyse complex research data
- the ability to develop and apply new research techniques to solve existing problems
- interpersonal and communication skills
- the ability to supervise technical staff.

So when the people at the KLATU Corporation look through the job applications the one thing that all applicants will have is an appropriate university qualification. What will set the applicants apart will be the ability of each to demonstrate their skills and abilities under each of the other selection criteria. Thus you can see the importance of all the things you learn at university.

This is just one example of the importance of graduate attributes, but the principle applies to most, if not all, professional positions, whether you are applying for a position as a physiotherapist, a teacher, a vet, or a mining engineer. Thus the goal of this book is to provide you with the means to develop essential academic and professional skills for use in science-based professions.

Transition to university

Going to university for the first time is much like starting a new job or moving to a new town. There is much that may be new to you at university, including the culture and the language of higher education. The types of problems encountered by students in any science-based course can be unique to that course, and to each student. However, many of the difficulties you face as a new student at university are common to most students, irrespective of whether you have just completed secondary school or have not studied for some time. Some of these are directly related to aspects of teaching and learning, but social, economic, personal, and health factors all have a part to play. On top of this is the demanding workload expected by the teaching staff, the pace and level of the work, the sometimes unclear demands and expectations of different topics, the differing nature of teaching methods, and learning the general conventions of the higher education system. Let us look at the nature of the differences between secondary and tertiary education (figure 1.1).

Figure 1.1   Differences between higher education and secondary school

Higher Education	Secondary School
Attendance in higher education is purely by choice.	All Australians undertake compulsory secondary education.
Higher education costs money, either in the form of HECS, or full fees.	Secondary education is nominally free but costs considerably more if you attend a private school.
The academic year is 26 weeks divided into two 13-week semesters, with assessments at the conclusion of each semester.	The school year is about 40 weeks long with classes extending over the whole year.
Class sizes can be very large with up to 1000 students, and you are just a 'face in the crowd' of a student population of 20 000 or more.	Classes generally have 30–40 students and you may be at the top of a school of 1000 or more students.
Each topic has a set curriculum but there are many things you need to learn that are outside the content of your topics.	You work to a rigid and set curriculum guided by your teachers and external examination authorities.
You need to work out your own timetable, find the time and place of your lectures, tutorials and other activities. Attendance in some parts of your course is optional.	You spend the large part of each day in a structured learning environment, with fixed classes and set breaks for lunch and recess.
You need to set your own priorities, monitor your progress, and become an independent learner.	You can rely on your teachers to remind you of your responsibilities and to guide your learning.
University is an environment where you are expected to take responsibility for your own actions and their consequences.	As a 'child' you are nurtured and protected by the school system which sets the tone in terms of moral and ethical requirements.
Handing in set work is entirely up to you; no one will check to see if you have done so. Feedback is mostly written so it is up to you to seek help or clarification.	Set work is regularly marked and checked with direct feedback provided by the teachers as to how well you are doing.

As the table indicates, perhaps the fundamental difference is the expectation that at university you are expected to behave like an independent adult and that you make decisions and take responsibility for your learning. It has been often said by students that university has a 'sink or swim' approach. This aspect of self-reliance is further reinforced by the following comments recently made by students about their university experience.

Students' experiences of university

I came with much anxiety to Australia to study, but this week (Orientation Week) has helped put some of my fears at ease. The system here requires very much more of students, where back home the teachers tell us exactly what to study and how to study it. I now know that I will need to learn in a new way to become more independent just like the local students if I am to be successful.

Rangit, first-year Chemistry

I am finding it really hard going at the moment. I like having someone around to tell me whether I'm doing the right thing or if I'm on the right track. I find this really helps me. Also, as a student doing three lab courses I seem to spend a lot of time doing things without knowing why I am doing them nor understanding what it is that I am actually doing.

Celia, first-year Biology

University study is very hard. It is very demanding and time consuming if you do all that is required of you. I think a good student needs to learn self discipline and organisation. There are so many things to get used to. I guess you have to rely on yourself. It's really like a process of self discovery isn't it? It would be very helpful though to get more guidance and clear explanations of course requirements.

Carla, third-year Biology

I like university so far, it's better than school, really big though compared to the school I went to. One of the things I miss compared to school is that we don't really have any proper discussion times. We just seem to be always busy doing things. It would be nice to have more small-group tutorials so we can actually ask questions and not just answer problems. It's all just lectures and labs and computer stuff but little opportunity to actually ask questions.

Nick, second-year Information Technology

One of the key things I have learned over the last three years is to not give up when things get difficult. You need to learn to make sacrifices for the sake of your study and to keep your mind focused on why you are here. Go to all your lectures and tutorials and get as much information from other students as possible. You have to learn to accept failure as part of success. And also remember to have a good time.

Beverley, third-year Nursing

It's all about dedication and working your butt off. The most important thing I have learned is to take responsibility for myself and work things out. You have to. The faculty staff have their own things to do so they cannot look after you. Once you realise it's all up to you it's not too bad.

Paul, third-year Engineering

If we were to look at the common areas of difficulty facing new students we could group them accordingly:

- academic preparedness
- academic progress
- financial affairs

- personal concerns
- family and social issues
- isolation
- uncertainty.

Let us take a look at these difficulties by referring to a range of questions asked by students in their first year of university, some of which you are likely to ask yourself.

Academic preparedness

- Are you in the course that was your first choice?
- Have you studied in this subject or topic previously in high school?
- Have you been away from study for a number of years?
- Are you 'burnt-out' after Year 12?
- Are your academic skills up to the requirements of higher education?
- Will you cope with the workload demands?

Academic progress

- Are you lagging behind in your course?
- Are you regularly late with assignments?
- Do you miss lectures, tutorials, or laboratory classes?
- Do you find the level of work required of you just too difficult?
- Is there simply too much to do?
- Is the approach to teaching and learning at university not what you expected?

Financial affairs

- Is money always a problem for you?
- Are your tuition fees due?
- Are you working too many hours in a part-time job?
- Do you have enough money for rent, food, and travel expenses?

Personal concerns

- Are you anxious, stressed, or not coping?
- Do you have relationship problems?
- Have you recently lost a friend, family member, or pet?

Family and social issues

- Is someone in your family getting you down?

- Is your partner not supportive enough?
- Are your children too demanding?
- Do you have a suitable place at home to study?

Isolation

- Do you feel all alone and confused?
- Do you feel like a little fish in a very big pond?
- Do you lack confidence to join university social or sports activities?

Uncertainty

- Are you concerned about your career and future job prospects?
- Are you anxious about all the questions above?
- What is the meaning of life, the universe, and everything?

These and many other difficulties have been faced by students in the past and will continue to face students into the future.

Balancing life and study

There is no simple answer to this issue. You have to work things out for yourself, but valuable allies are at hand. Family, friends, partners, and fellow students, as well as lecturers, tutors, and student agencies on campus, are all sources of support and assistance to help you find a solution to the difficulties that will inevitably arise (figure 1.2).

Figure 1.2   Representation of the three broad aspects of your life that may impact on your studies

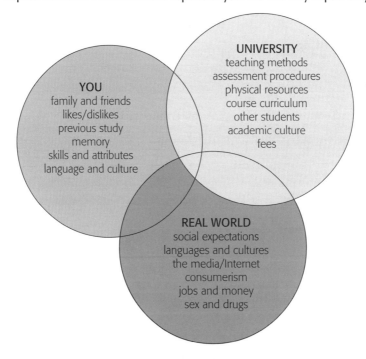

Studying is a balancing act where you have to adjust your lifestyle to enable you to do things that you want to do as an individual, as well as the things you need to do to fulfil the requirements of university and your life outside of study. This can be represented by the three discrete areas in the diagram above: you and your life, the world of the university, and the larger world. What you need to do is to find the right balance between the different facets of your life. Unfortunately there is no easy way for doing this, and some of it will be a matter of trial and error. One of the key strategies in balancing your commitments is to have a systematic approach to the management of your time, and this will be the subject of the next chapter.

Skills for academic success

So a key question you may be asking at this stage may be 'What are the skills required by students to succeed at the tertiary level?' To look at a possible answer to this question, two groups of students in science topics (first-year and third-year students) were asked to list the qualities they believed to be most important to them as students in a science-based course. The responses were collated and put into one of ten categories. Figure 1.3 shows the student rankings of the top five listed categories.

Figure 1.3   Rankings by first-year and third-year science students of the qualities and skills for successful study

Student factors for success	Third-year students (N = 92)	First-year students (N = 132)
Motivational aspects	1	1
Learning skills and strategies	2	4
Work ethic and self-discipline	3	3
Self monitoring of learning	4	5
Things about study that make you feel good/bad	5	2

The most important message to take from this data is that both student groups list the same top five factors for success, and that of these motivation is the key factor that students think contributes to their success. Having a high work ethic ranked third for both groups, and the third-year students placed a greater value on learning skills and strategies.

Recommendations

All students commencing university have some difficulties adjusting to tertiary study. This is a natural part of the growth and development of each of you as human beings. We hope that the challenges you meet during the course of your academic life will enrich you and lead to educational, vocational, and personal development.

The following is a list of strategies that may help you overcome, or at least lessen, some of the difficulties associated with being a commencing student.

GENERAL TIPS

- Maintain consistent and conscientious effort and motivation.
- Stay up-to-date with set tasks.
- Establish a routine that suits your study needs and your course requirements.
- Seek assistance early when and if you need help.

SPECIFIC TIPS

- Attend any academic orientation programs and relevant preliminary lectures.
- Form a peer study group, particularly for those subjects you find difficult.
- Familiarise yourself with the university and *all* of its resources.
- Seek clarification of expectations, requirements, and assessment procedures.
- Know when and where to find your lecturers and tutors.
- Find out where else to seek help: counsellors, study advisers, or senior students.
- Have a flexible study plan that allows for leisure and social activities.
- Learn to ask questions, like 'What should I do next?' or 'Who should I ask?'
- Do all the compulsory work required of you and hand it in on time.
- Be prepared for lectures, tutorials, and laboratories by reading ahead.
- Learn how to effectively use all the resources available to you.
- Do something straightaway when problems arise, don't just leave it.

Summary

Like all new and unfamiliar environments, university at first can seem a large, impersonal, and unfriendly place. If you are a student from overseas, interstate, or a rural community, you may be among those who feel most isolated. You may feel homesick and sometimes fail to see the relevance of your study tasks. At times, a normally able and successful student may lose interest and become disheartened with a particular topic or course. So why bother? Keep in mind the following:

- You are paying for your education, so get the most out of it.
- Success or failure is largely dependent on you.
- Study can be fun and enjoyable. It beats many of the options.
- You are studying for a purpose, so keep that in mind.
- Think about tomorrow, but act today.

2

Being Organised

Things that matter most must never be at the mercy of things that matter least.

Johann Wolfgang Goethe, German polymath (1749–1832)

Key Concepts

- Why time management is important
- Setting goals and priorities
- Planning for study success
- Developing a study timetable
- Some things that waste time
- How to overcome procrastination
- Time management strategies

Introduction

How many times have you said, *'If I only had more time!'* More time to spend with family and friends. More time to socialise or play sport. Maybe more time to read a good book or watch movies. Or perhaps you would like more time to think about what you would do if you had more time. In today's society where most of us spend a great deal of energy rushing around, it may seem as though there are just not enough hours in the day to get everything done. It is an ironic tragedy of our modern lifestyle that the more time-saving gadgets we have, the less real time we seem to have to enjoy our extra time. This chapter will assist you to develop better strategies for the management of your time and your study tasks.

Why time management is important

One of the first things you learn as a tertiary student is that studying requires a commitment in time and that you may have to make personal sacrifices in order to succeed. So a question worth considering at this point might be: 'Is effective time management an important factor in successful

study?' Intuitively, most of us would probably answer yes, but is it? To add some substance
to this, consider the following study on time management for tertiary students in the USA.

Time management

A study of college students in the USA looked at the relationship between time management and
academic success as determined by their grade point average (GPA). Students were asked to complete a
survey of three aspects of their time management:

- T1 = short-term planning
- T2 = attitude towards the use of time
- T3 = long-term planning.

At the end of their four-year college course, students' GPA was correlated to measurements of these
three aspects of time management. The results showed that the first two of the time management factors
had a positive relationship to academic success, and, further, that these correlations were larger than the
correlation between GPA and the traditional guide to academic success, the scholastic aptitude test (SAT).
The important message is that time management matters.

Journal of Educational Psychology, vol. 83(3), pp. 405–10, 1991

Studying in any of the science-based areas generally means that your days on campus are
structured, with set times for lectures, tutorials, and practical sessions. You may also have off-
campus activities such as field trips or visits to professional practitioners. On average, you can
expect to have about 20 to 25 hours of formal contact time per week. In addition to these
hours you will need to allocate extra time to work directly on your studies, as well as for
other commitments. So a key question you are likely to ask yourself early in your studies is:

How do you cope with all the demands on your time?

First, you need to be well organised, and, second, you need to be aware of the factors that
most impact on how you allocate your study time, which may depend on:

- your characteristics: personal, physical, and emotional
- your preferred mode of studying: time, place, and duration
- other people: family, friends, partners, social contacts
- your commitments outside of study: work, recreational, social, and sporting.

With these factors in mind, you need to prioritise different aspects of your life so that
you can most effectively use your available time. The following parts have been designed
to assist you.

Setting your goals

There is always time for whatever you consider to be important. It is a matter of setting
priorities. How many times have you said, 'I don't have time.' What this usually means is,
'I consider other things more important.'

So your first task is to sort out your priorities, and to see if and how these priorities are linked to the reasons you choose to spend several years studying. Use figure 2.1 to rank your priorities according to their level of importance.

Figure 2.1   Your priorities and their importance

Ranking	Your priority areas
Vital	
Important	
Desirable	
Neutral	
Not important	

Now consider these priorities in terms of your goals, keeping in mind that goals and priorities are not necessarily the same, and that they may change with time and circumstances. Write down specific long-term and short-term goals.

- Where do you want to be in three years from now?
- What are your goals for this year?
- Make a list of what you hope to achieve each week or month. Rank each of them in order of importance and tick them off as you achieve them.
- At the beginning of each study period, set goals for that session.

Planning

Having determined your goals, both long-term and short-term, you need to develop a set of appropriate strategies to attain them.

1 Overall study plan

Perhaps the first part of planning your studies is to have an overview of your course as a whole. In some courses, each topic you study at each level is prescribed and is thus a compulsory topic. In most courses, however, there are *core* and *non-core* topics. For example, you may wish to study marine ecology, which is a three-year science degree. Look at the topics you need to study in the third year and then see which second-year topics are prerequisites for these. Then do the same for second-year topics. Be selective in your choice of non-core topics and select those that give you the greatest scope for subsequent years, or topics in which you have a particular interest. A little thinking ahead early on will ensure you can complete your studies in the minimum time thus saving both time and money. Finally, also be aware than when choosing non-core topics, these can have timetable clashes with core topics, so choose carefully.

2   **Semester schedule**

Having a basic outline of each semester's main requirements provides an overview of each semester. It lets you see which weeks are full and which ones offer greater flexibility. Construct your schedule to include all major events, exams, tests, and due dates for assignments.

3   **Weekly schedule**

Having a weekly schedule lets you plan the details of each week. You can select the best times to study, exercise, work, and relax. Such a schedule helps eliminate the constant need to decide between 'must do' and 'want to' activities.

4   **Daily schedule**

A daily 'things-to-do' list is a simple and practical way of ensuring that what you need to do is done. The best time to prepare this schedule may be at the end of each day.

Aspects of university study and the factors outside of study may have different degrees of importance and impact differently on your academic success. Some activities have fixed time requirements and some are more flexible, so try to determine which category they belong to. For example, we can differentiate between a number of tasks and activities based on their flexibility as follows:

- Fixed activities: classes, work commitments, travel, sport
- Flexible activities; sleep, recreation, study, relaxation, socialising, reading.

The study timetable

One of the most commonly used planning approaches is to have a study timetable to keep track of your weekly commitments and activities. Most student diaries have a timetable with hourly time-slots for each day of the week. When filling in the weekly planner include the following:

1   Fill in all your fixed commitments: lectures, tutorials, and practical sessions.

2   Include the essentials of your daily routine, such as time with family and friends, travel, shopping, exercise, and social and recreational activities.

3   Be specific when organising time for certain tasks; for example, do not have a block of time for 'study'.

4   Be flexible enough to allow for changes in commitments and for tasks to take longer than anticipated.

Exercise: developing a timetable

Let us develop a weekly timetable for Sam, a first-year science student, who is studying marine biology with semester 1 subjects chemistry, biology, earth science, and computer science. Sam plays soccer and is the captain of the local team, the Walliroos, attends practice on two nights each week, and plays matches on Saturday afternoon. Sam relies on the bus for transport and lives in a flat which is 10 km from the university, and prefers to study at night with 23 hours of formal class time per week. In between classes Sam has some one-hour or two-hour time slots to spend on low-level study activities, such as looking for materials in the library, using the computer labs, preparing for assignments, or catching up on required reading. Sam has timetabled an extra 20 hours of studying at home, to prepare for laboratories (Labprep), write up assignments (Assign), complete laboratory reports (LabRep), review lecture notes (LectRev), and prepare for tutorials (TutPrep). Sam has also included a commitment to working at McBurgers (WORK) and sport (SOCCER), and has added these to the timetable in figure 2.2.

Figure 2.2   Sam's weekly timetable

	Monday	Tuesday	Wednesday	Thursday	Friday	Saturday	Sunday
8–9 am	Travel		Travel		Travel		
9–10 am	ChemLect	Travel	ChemLect	Travel	ChemLect		TutPrep
10–11	EScLect	BiolLab	EScLect	ChemTut	EScLect		TutPrep
11–12		BiolLab					
12–1	BiolLect	BiolLab	BiolLect	BiolLect			WORK
1–2							WORK
2–3	CompLect	EScLab	CompLab	CompLab	ChemLab	SOCCER	WORK
3–4		EScLab	CompLab		ChemLab	SOCCER	
4–5	Travel	EScLab	Travel	Travel	ChemLab	SOCCER	
5–6	SOCCER	Travel			Travel		
6–7	SOCCER			SOCCER	WORK		
7–8	Assign	LabRep	Assign	SOCCER	WORK		LectRev
8–10	Labprep	LabRep	Assign	LectRev	WORK		LectRev
10–12	Labprep	Assign		LectRev			

When you develop your own study timetable, you should keep in mind the following:

1   Are you an early bird, or a night owl? Use time slots when you concentrate best for the most demanding tasks.

2   Do you have time slots between set activities? Use these for work not requiring you to be at your intellectual best, such as photocopying, library searches, or organising lecture notes.

3   Do you need to allocate more time for certain activities? Some things take longer than you may have planned, so allow for flexibility.

4   Are you up to date in all your subjects? Revise your schedule on a regular basis.

5   Are you devoting too little or too much time to various tasks? Research indicates that short bursts of study of up to two hours are more effective than sustained periods without a break.

Resources

Something directly related to making the most effective use of your time is making the most of resources available to you. So when you are about to start on a new aspect of study, a new assignment, or write up a laboratory experiment, keep in mind the resources available. A few minutes consulting the most suitable resource may clarify problems and save time. This is one way to ensure the small problems of today do not become the large problems of tomorrow.

Resources available to assist with your studies include:

- family and friends
- classmates and other students
- teaching staff: lecturers, tutors, demonstrators
- course materials: booklets, lecture notes, handouts
- text books, periodicals, assignments, tests, problem sheets
- electronic sources: Internet, telephone/mobile/SMS, television, radio
- student academic advisers and counsellors, course advisers, student associations.

Using your time well

It is essential that you structure your lifestyle so that enough study time is available to suit your study habits. This may require a lifestyle change, especially if you are living with others (family, friends, partner, or housemates) who are not students and may not understand your study needs. Living with parents can be an advantage, but is not always the best option. Students who are married and have family responsibilities will be those who may need to make the biggest sacrifices and changes to their routines. If you live away from home for an extended period for the first time you may find your new life challenging and difficult. For example, you may be living on campus. If you are on your own for the first time, socialise with other students and get to know them. Your classmates are often your best study resource as you share similar experiences.

The following are suggestions from students, which may help you organise your time and your studies.

TIPS FROM SUCCESSFUL STUDENTS

- Find a regular study place that can be safely left and returned to.
- Ensure there are as few distractions as possible when you are studying, and that the study atmosphere suits your needs.

- Count the hours spent on study in a 'good' or 'typical' week, so that you can identify what is realistic for you.
- Make use of 'hidden' time, such as having a shower or preparing a meal, to call to mind a recent lecture or an essay you are working on.
- Review course content materials regularly and often. Reviews should be cumulative, covering briefly all work done.
- Develop a filing system to organise your notes and handouts, so that you do not waste time searching for them when needed.
- Start assignments while your memory of the assignment requirements is fresh.
- Limit your blocks of study time to no more than two hours on any one topic. Your ability to concentrate decreases quite rapidly after about one and a half hours.
- Take a break from each topic and study something else. This will help keep your mind fresh to maximise your efficiency.
- List your tasks according to their priority. Do the most important ones first.
- Overestimate your time. A good rule is to estimate how much time you think something will take and then double it. If this is too generous, you will have some free time for other tasks.

Late submissions

Different universities and departments have varying policies regarding extensions and late submissions for assignments. Some departments will not accept work submitted after the due date, while others will penalise you for each day late. These policies are usually set out in course handbooks. If you are falling behind, talk to someone who may be able to help you, such as a lecturer or study skills adviser. If you are sick, a medical certificate may be required. If you are consistently late with assignments, is it for one of the following reasons listed in figure 2.3?

Time wasting

There are many ways in which time may be wasted. Some factors may be beyond your control, but you may deliberately waste time in order to avoid unpleasant but necessary tasks. It is important to be aware of the factors that impact negatively on your study regime. Of course, having identified which of the factors tend to impact on you most, it is another matter to address these factors. You need also to accept that there will be external factors which impact on you and which you are powerless to control. Look at the list in figure 2.4 and identify how many of these apply to you: all of the time, regularly, some of the time, or hardly ever.

Figure 2.3   Problems and possible solutions for late submissions

Problem	Possible solution
Other things have higher priority.	Reconsider your priorities: have they changed? Has study become a lower priority?
You misjudge how long tasks take.	Keep a record of the time spent on each assignment. Use this as a guide when planning your next one, making allowances for degree of difficulty.
Due dates suddenly appear and take you by surprise.	Do you have a study plan? Break assignments into small steps and set deadlines for each step.
You think there is a 'right' answer but you cannot find it.	There is not always a 'right' answer, unless you are referring to a numerical answer to a specific calculation.
You are not sure if you are doing what is required.	Read all the instructions to ensure you are doing what is required. If in doubt ask.
There is too much information.	Start small. Eliminate all but the essentials, then build gradually from there.
You cannot get started.	Talk to other students or your lecturers. Just start somewhere; put something down in writing.

Figure 2.4   Factors which impact on your time

Factors outside your control	Self-imposed factors
Travel problems: car, bus, train, bike	Anxiety over what may never happen
Emergencies	Computer games, TV, email, Web surfing
Family issues	Mobile phone and TXT messages
Illness or injury	Dreaming about winning lotto
Library is too busy	Lack of concentration
Cannot get onto a computer	Lack of planning or poor strategies
Other people's problems	Not reading the instructions
Interruptions	Poor or inappropriate academic skills
Problems with your partner or flatmates	Socialising with friends or family
Unclear instructions for assignments	Procrastination
Too many demanding people	Weather is just too nice to study

Procrastination

For the list of self-imposed factors above, perhaps the overriding problem is that of procrastination. Procrastination means delaying or avoiding something that must be done, and some of us develop this to an art form. The key to overcoming procrastination is to be decisive. Make decisions that will lead to positive outcomes rather than wasting time and energy thinking about them. The following strategies will help overcome a tendency to procrastinate:

1   Recognise that procrastination is a habit that cannot be changed unless *you* want to change it.

2   Be motivated to change. Self-belief is required to change habits.

3   Start small. Large goals are often difficult to reach and require concerted effort and time. Small, achievable goals are the best way to change your habits.

4   Keep going once small goals are reached. Setting and reaching goals becomes an ongoing process to success.

5   Failure is a part of success. Learn to accept that you cannot always achieve the ideal.

6   Reflect on the reasons for failure and use those insights to improve your next goal-setting session.

TEN TIPS ON BETTER TIME MANAGEMENT

There are many strategies used by successful students. The following is a list of some that may help you to be a successful time manager. Keep in mind that the key factor in all forms of management is attitude.

1   Set clear objectives and priorities and recognise the difference. For example, your objective may be to obtain a tertiary qualification to be a vet, but your priority is to maintain your health and fitness in the process.

2   In order to achieve your objectives you must decide which are the most effective activities that will help you achieve those goals. Remember that some tasks are a means to an end and some are an end in themselves.

3   Plan your time and your study. Have a long-term plan, a short-term plan, and a daily 'to-do list' in order of priority.

4   Stay positive. You can do this by making sure that during each day you accomplish something worthwhile and reward yourself for it.

5   Accept that you will make mistakes, learn from them, and don't repeat them. In planning your activities allow enough flexibility to accommodate mistakes, mishaps, and the unexpected.

6   Use all information sources so that you have the best possible chance of doing most things right the first time. In that way you won't have to waste time repeating things.

7   Finish what you start. Some people (such as Leonardo da Vinci) jump from one activity to another until their objectives have been met. Others (such as Marcel Proust) tend to focus for extended periods on a single activity until it is complete, and only then will they move on to the next activity. It does not matter what approach you use as long as it works for you and you get things done.

8   Conquer procrastination. Be honest with yourself and eliminate or minimise time-wasting activities.

9   Accept that you cannot avoid the unexpected, so learn to handle things when they go wrong.

10   Enjoy university; it is a rewarding experience.

Summary

Studying requires good organisational and management strategies; you will be effective by:

* being less rushed and have more time
* keeping motivated to succeed
* minimising the opportunities for procrastination and avoidance
* avoiding meeting deadlines at the last minute
* reducing your stress and anxiety levels to perform at your best.

A quote from Stephen Covey's book *Seven Habits of Highly Effective People* may put things in perspective:

> The successful person has the habit of doing the things failures don't like to do. They don't like doing them either … but their dislike is subordinate to the strength of their purpose.

PART 2

Learning and
Researching

3

Modes of Learning for Higher Education

I am always doing that which I cannot do, in order that I may learn how to do it.

Pablo Picasso, Spanish artist (1881–1973)

Key Concepts

- What is learning?
- How your memory works
- What is a learning curve?
- Finding your preferred learning environment

- Learning for meaning
- Monitoring your own learning
- Tips to improve your learning outcomes

Introduction

We learn all our lives. As children we learn to walk and talk. We learn to play games, from simple ones like skipping rope to complex games of strategy like chess. Most of us learn how to ride a bike and to play a musical instrument. In the academic context, we also learn from an early age, like the letters of the alphabet and how to use numbers. From our early school years we also become aware that learning appears easier for some people than for others, who seem able to read or hear something and understand it immediately. Unfortunately, most of us have to work harder. If you are new to tertiary education, the learning environment may be very different to your experiences at high school. Chapter 1 looked at some of these key differences, the greatest being the expectation placed on you to be an independent learner.

This chapter has two parts. The first is basic information on the different modes of learning suited to higher education. The main focus is to assist you to recognise those traits that apply to you. The second part is more pragmatic and offers some hints from former students on how to improve your study habits. The premise underpinning this chapter is that if you know more about the learning process, and in particular about your *own learning style*, this will assist you to become a better learner.

What is learning?

Learning is the process of acquiring new skills and knowledge. All of us are different, so we all have our own methods of learning and studying. You may know someone who appears academically more successful, yet they seem to have more free time than you do. How is that possible? We all have our own unique learning style and rate at which we learn and so you must work out a system that meets your needs and not be swayed by how others study, though learning from the success of others is important.

Your personality affects how you experience the world and how you learn. There are a number of qualities, skills, and experiences you bring to university that may affect how you learn, including:

- intellectual abilities and skills
- cultural and language background
- prior learning experiences
- preferred learning environment
- orientation to learning
- personality traits
- prior knowledge
- moral, ethical, philosophical, and spiritual values
- career expectations
- memory skills.

Science-based courses are often dominated by content, so that much of what you do on a daily basis at university involves listening to lectures, reading text books, taking notes, making summaries, completing set tasks, writing-up laboratory results, and solving tutorial problems.

In order to maximise your learning, you need to engage your mind at the high end of the learning scale. Higher learning takes place when you design, plan, synthesise, extrapolate, critique, hypothesise, and evaluate. To ensure that you maximise your potential for higher learning it is vitally important that two things occur. First, from a learning perspective, it is vital that you ask questions of yourself, your fellow students, and your teachers as a guide to your understanding of key concepts and your ability to use those concepts. Such questions might be: Why is this important? How does it work? How can I use this information? Can I extrapolate from specific examples to generalities?

Second, from a teaching perspective, it is equally vital that assignments, tutorial problems, and practical work with which you engage cover the range of higher cognitive levels and not just the superficial level of 'recognise, describe, or list' which test only at the recall level. Good examples of assignments and study tasks will demand that you analyse data or information, draw conclusions, extrapolate or make predictions based on your conclusions, and even design an experiment to test an hypothesis.

Your role as a student is to assimilate information and become familiar with it so you can construct new knowledge that becomes part of your existing knowledge when next you approach an unfamiliar area. The key is to strive for understanding so that you are able to use new knowledge, and one of the key tools you use to to so is your memory.

Memory

Memory is one of the most important tools for learning and understanding. It may seem unfair that some people have a much better memory than you do. Does this mean they are smarter? Does it mean they have some advantage when studying? The answer to the first question is, no, they are not necessarily smarter. For example, some people with autism often have a unique ability to recall or process information accurately and quickly, like the character of Raymond in the film *Rainman*. These people are called savants. A question may be, '*What day of the week was it on 1 April 1827?*' The savant may know the answer to the question, but the answer for them is probably meaningless. As to the second question, the answer would be a definite yes: a good memory does help studying. Being able to easily recall information allows you to spend more of your cognitive space, not to mention more of your time, on higher order cognitive tasks.

So how does memory work and what can you do to improve it? Figure 3.1 shows a schematic representation of memory, which has both a short-term memory (STM) and a long-term memory (LTM) component. In a learning situation, you are exposed to a variety of learning stimuli. You hear a lecture, you see words on a page, you feel and touch equipment in the computer suite, and you smell the reagents in a chemistry laboratory. All of the experiences detected through your senses leave an impression in your memory, but only a small proportion leave a lasting impression.

Each of your sensory experiences gets loaded into your STM and stored in such a way that you are able to recall it. If, however, you do not recall that information or use it in some way, about half of it becomes 'lost' in about 24 hours. In fact, most of the things you see, hear, or experience during a typical day are lost as they are considered as unimportant. In order to transfer your STM information into your LTM you need to *revisit* the information on a regular basis. Think of trying to remember an important telephone number. You write it down. You read it out loud to yourself, probably several times. The next day you recall the number and dial it. You do the same thing a couple of days later. Perhaps a few days or a week later you do the same thing. Before you know it,

Figure 3.1   Schematic view of memory

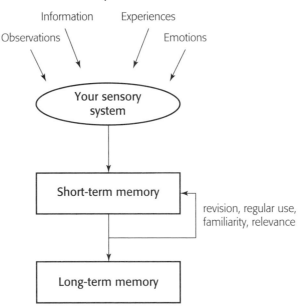

that number has become part of your memory; you can now dial it without thinking, and it becomes automatic. Learning is exactly the same process. You have to become familiar with new information so that it becomes a fixed part of your memory. This can only happen through continued practice and rehearsal.

The diagram above could imply that there are two different memory centres, one for STM and one for LTM. This is not the case. The time that you retain something in your memory is more of a function of how well you have learnt the material in the first place, so it is more to do with depth of processing. Memory is also linked to relevance, familiarity, and the simplicity or difficulty of the material. Two examples will illustrate this.

Look very briefly at the two words below, each of which has about the same number of letters. Which one is easier to remember? Why?

- Supercalifragilisticexpialadocious
- MethylmalonylCoaminetransferase

Now look at the list of country names below, again as quickly as possible, and see which list is easier to recall.

- China, Japan, Mali, Malta, Nepal
- Guinea-Bissau, Azerbaijan, Turkmenistan, Burkina Fasso, Bosnia Herzegovenia.

All sciences are rich in complex names, terminology, and technical details, and you need to remember much of this information. Thus one of the key skills that will help you to improve your learning is to improve your memory.

TIPS TO IMPROVE YOUR MEMORY

1   Look for patterns, such as visual or aural patterns.

2   Use acronyms or devise mnemonics for related information, such as BEDMAS for Better Excuse Dear Mister Angus Smith for the order of the mathematical operators Brackets, Exponents, Division, Multiplication, Addition, and Subtraction.

3   Set new information to a tune that you like or which is easy to remember.

4   Draw diagrams or flow charts.

5   Make memory prompt cards.

6   Revisit information often.

The learning curve

How many times have you heard the phrase, 'They are on a steep learning curve?' The term learning curve refers to the rate at which learning takes place and is used in all sorts of contexts, yet it is rarely used to describe the process of studying. Figure 3.2 represents a learning curve, where the horizontal axis represents time and the vertical axis represents the degree of learning and comprehension. When you become familiar with new information and are able to recall and use it effectively, you may be able to progress to a higher level of understanding, that is, you are suddenly aware of how something works, or you understand something that you previously found difficult.

Figure 3.2   An example of a learning curve

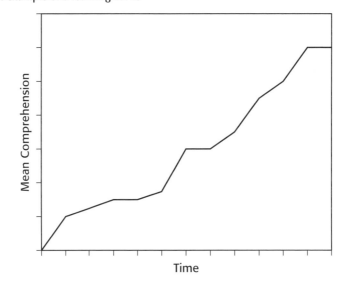

Learning is not a linear process. It is full of bumps and curves. Sometimes you go forwards, sometimes you stand still, and at other times you appear to go backwards. Time, effort, and perseverance will help you get through the times when nothing makes sense and when it all clicks into place.

So how do you optimise your learning? Optimum learning takes place when you are sufficiently stimulated to learn. The type and level of stimulation is different for everyone but similar in that it is a mixture of positives and negatives.

Imagine you are 10 years old and it is Christmas Eve. You have been looking forward to it because you are keen to learn to play guitar and are hoping for a guitar as a present. On Christmas morning you get up early and open the large present your parents have given you to find that they have bought you ... a violin! Now just imagine how stimulated you are to learn to play the violin.

So when it comes to learning in the academic context look at the terms below (figure 3.3) and ask yourself which of these positive and negative stimuli apply to you and to your learning success: sometimes, all the time, hardly ever, or never?

Figure 3.3   Optimising your learning through positive and negative stimuli

Positive stimuli	Negative stimuli
Excitement	Anxiety
Motivation	Stress
Good results	Poor results
Study partners	Fear of failing
Course interest	Boring topic
Look forward to the labs	Can't understand the labs
Like being busy	Too much to do

There are times when your motivation levels can be a bit low and the negative aspects can be high, particularly when doing things that you do not like. A way to help overcome this is, first, to keep focused on the end result: the end of the topic, end of semester, or end of the course.

Second, you have a choice as to how to plan your schedule to incorporate different learning needs. Do the routine boring stuff at your low times. When the demand of the task is high, do such tasks at your optimal time. Reward yourself for good effort or a good result in a test or assignment. Lastly, avoid negative thoughts that can increase your anxiety and stress. Do not focus on the things that cause resistance to your study, but seek solutions (figure 3.4).

Figure 3.4   Strategies to aid motivation

Problem	Solution
The lecturers in this topic are hopeless.	Not all lecturers are created equal. Ultimately, your success or failure depends on you.
The library is always too crowded when I go there.	Look for other times when it is less busy, such as first thing in the morning, lunchtime, or late afternoon.
Why do I have to do this stupid assignment?	All assignments are designed to assist you develop skills and knowledge, even when it may not be obvious to you.
Everybody finds this topic really easy except me.	This is unlikely to be true. Speak to other students and get their views.
I will never get this assignment done on time.	Why are you late? You may need to rethink your schedule and plan ahead better.

Preferences for learning

Your learning environment can have a considerable impact on your productivity, so one of the first steps to finding out what type of learner you are, and thus enable you to maximise your productivity, is to find your preferred physical learning environment. In order to do so, you should ask yourself:

What is my preferred learning environment?

Do you prefer the lighting to be	bright	subdued	low
Do you prefer to study	alone	with a friend	in a group
Do you prefer to study in the	morning	afternoon	night
Do you prefer to study	at home	in the library	elsewhere
When you study do you prefer it	warm	cool	no choice
Do you prefer to study	in silence	with music	watching TV
When you study do you have	snacks	tea or coffee	chocolates
Do you prefer to study	in short bursts	for 2–3 hours	for sustained periods

Establish your preferred studying options, then stick to what works best for you. Once you establish a routine you will be more productive.

Sensory aspects of learning styles

A preferred learning style has been described as being the consistent way in which you perceive, conceptualise, and organise information, and how you set about your learning tasks. A particular learning style is related to your personality, and though a learning style is relatively stable over time, it can change. One aspect of how people learn is based on sensory perceptions. Look at the three categories below and see if you can recognise those traits that you see in your learning style:

Visual

If you have a preference for a *visual* learning style, you

* seek visual cues, such as facial expressions or gestures
* use diagrams or other visual representations
* recognise words by sight
* look for patterns
* remember things by visualising them.

Auditory

If you have a preference for an *auditory* learning style, you

* respond to verbal instructions
* learn concepts and solve problems through discussion and debate
* use sound and rhythm as memory aids.

Kinaesthetic or tactile

If you have a preference for a *kinaesthetic* or *tactile* learning style, you

* respond to active participation
* use physical and practical aids for remembering
* like hands-on projects and group activities.

  Few of us fit neatly into only one of these learning styles. The key is to find which of these suits your learning, and when you recognise those traits in yourself, use that knowledge to build on and develop skills to improve your learning.

Learning for meaning

The sciences are rich in information, with many facts, figures, relationships, and terminology. Remembering information is not enough if you cannot apply it to specific problems. One of the best ways of learning is memorising. If the sole purpose of memorising information is to be able to regurgitate it in a test or exam, then you may not have learnt very much. The purpose of learning is to be able to do something with your new knowledge, in line with the higher order cognitive levels expected of university graduates mentioned earlier. So memorisation can be a 'tool' for better learning if you ask yourself questions such as: Why is this important? How can I apply this information? Can I see links between this and previous course content I have learned?

As an example, say your have learnt how to use a particular method to tackle a genetics problem by going step-wise through examples and then learning that method by heart. If you are presented with the same type of problem in an exam, you should be able to answer it. If, however, the examiner decides to set a different type of problem, or describes the problem in a different way, you may not know where to start developing an answer or be able to accommodate the need for a different approach. So the key is to be flexible and to strive for understanding.

Monitoring your own learning

One of the key strategies that can assist you to become a better learner is to become better at actively monitoring, regulating, and evaluating your cognitive or learning processes. This form of self-regulation is a process *above* the ordinary cognitive levels and hence is referred to as *metacognition*. Students who are more able to monitor and evaluate their own learning and are more aware of how they learn tend to be more successful at the tertiary level.

The following describes the stages in the metacognitive approach to learning. To some extent we all do this intuitively, so perhaps it is a matter of fine-tuning how you do it.

TIPS TO MONITOR YOUR OWN LEARNING

1   Identify the nature of the study task.

2   Set target goals with specific desired outcomes.

3   Devise a plan to meet those goals.

4   Engage the appropriate cognitive skills: reading, writing, solving problems.

5   Monitor your activities progressively to see if you are succeeding.

6   Evaluate your outcomes in terms of the attainment of target goals.

7   Reflect on your achievements or perhaps lack of them.

8   Act on those reflections by modifying your approach for next time.

Advice from successful students

The following tips and advice are based directly on comments made by students who have been successful in their science-based courses, so here is an opportunity to learn something from the experience of others.

Keep up

The most useful advice is to stay up with or ahead of the class. If you fall significantly behind in science classes, it can be hard to catch up. Science courses are information-rich with facts, terminology, and relationships that need to be learned and understood as these will be assumed in later parts of the course. Fluency in basic course material is necessary to understand and learn the more complex ideas in the course.

The topic syllabus

Most courses have a guide that details essential information about the course, the content, evaluation procedures, and assessment expectations. Within this course guide you should also find a syllabus, which will provide a description of the course content and what you need to do in order to do well. Refer to the syllabus often to check that you are up to date with the course content, and that you know what is coming up next so that you can prepare *before* your lectures.

Get to know your classmates

Get to know some of your classmates in the lecture, laboratory, or tutorial classes. Exchange your phone number or email address with these students and keep theirs in your student diary. Your fellow students can be valuable allies and resources when you need information about a missed class or if you get stuck on problems.

Survey the entire work

Before you start learning for a particular topic or a given task, read through the major headings and summaries of the relevant chapters in the textbook, or download lecture notes if these are available. This gives you an overview of the 'big picture', and may assist in the selection of material to which you should pay particular attention while reading. Research shows that this simple step is one of the most powerful tools for learning, and that students who do this consistently achieve higher grades.

Reading, highlighting, and taking notes

As you read new material in textbooks or printed lecture notes, you should make written notes and highlight key words, phrases, terminology, or diagrams. On your writing page, initially use only one side of the page, for example the left half of the page. Transfer to the right side of the paper comments the lecturer made about the material during the lecture. You should end up with a set of lecture notes supplemented and complemented by information from your reading.

Highlighting is a skill that is very important if done properly and judicially. The tools of highlighting vary according to your own preferences, but the most popular are highlighter pens that come in a variety of colours. At first you may find the tendency to highlight too much material. In time, as you become more analytical and astute, you will notice a reduction in your highlighting as your skill in identifying the most important concepts improves. You may find it useful to use different colours for different levels of importance; for example use red for extremely important material and yellow for moderately important material. The process of reading and deciding if the material is important enough to highlight increases your memory of that material. It is the decision making and thinking that creates the memory.

The card system

Part of learning science is the simple but often derided skill of recall. Whether we like it or not, in every area of study there are certain things that we just need to remember. One simple way of storing and categorising important information is to keep a card system. Make up some cards, say 10 cm × 15 cm, and write important information from lectures and textbooks on the cards. It also helps to have different colours for different levels of information, with key concepts, perhaps in red, and supplementary material in another colour. Use one card for each major idea or theory, and include terminology, relationships, relevant factors, and equations. Figure 3.5 shows an example of a card for buffer solutions.

The biggest problem with textbooks and lecture notes is that it is often difficult to separate or find the must-know material. Because of this, you can waste hours finding the essential information. This is where the card system works well.

Figure 3.5   An example of a study card system

Front of card

ACIDS & BASES   card 8
Buffers and pH

Back of card

1. Buffers are solutions containing a weak acid and its conjugate base.   i.e. HA and A$^-$
2. Buffers have the ability to resist changes in pH.
3. The pH of a buffer solution can be determined by using the *Henderson-Hasselbach* equation.
4. $pH = pKa - \log \dfrac{[acid]}{[anion]}$

The key words and terms marked in red and blue alerts you to the fact that this is extremely important material. Writing the material helps to store the information. On the back of the cards write a definition about the material on the front. Also, number the cards so you can put them in order. As a way of testing your recall, turn the cards over and try to remember the information. Once you are able to regularly and accurately recall the information, you can place the card into the 'I know this material' stack. Now continue working on the material that you don't know until you can recall the material on all the cards.

Review information

We forget most of the things we hear during the course of a normal day. For example, by the next day at least half of what we heard in a lecture is forgotten, and if we wait another day or two, then about 90% is forgotten. To address this problem we need to revisit information on a regular basis. It is important to review your lecture notes regularly, and one simple way to do this is to reread the material written in your notebook. Work through the material in your notes paying particular attention to the material underlined. As you reread the material, highlight and star the material you believe is most important. Now reread the material you have highlighted or starred and high-speed review the material on the cards. Frequent reviewing like this is the best method to improve your recall of essential material.

TIPS FOR BETTER LEARNING

- Keep up with your course material as it is difficult to catch up if you fall behind.
- Know which material (terminology, equations, laws, and relationships) you need to commit to memory.
- Strive to understand key concepts before continuing. Knowledge builds so that each new concept and major idea relies on your understanding of previous concepts.
- Know your learning style in terms of strengths and weaknesses, study environment, and the style/approach that works best for you.
- Attempt all the required assignments, problems, and quizzes for your topics as they often give an idea of the level of understanding required.
- Find a place to study that suits your needs.
- Find a consistent and regular time to study.
- Review lecture notes and other course materials regularly and often.
- Be flexible enough to adapt and modify your study patterns and approach to suit the course materials or study topic.

Summary

Learning is a very personal experience and one that is influenced by your personality, your reasons for learning, your level of motivation, and your own learning preferences. Learning rarely progresses smoothly in a linear fashion and there are many challenges along the way. The more you are aware of what works for you, the better you should be at learning new things. Finally, science is rich in terminology, facts, and figures, much of which needs to be committed to memory. But remembering all this information is only the first step towards understanding its meaning and being able to apply the information to new situations.

4

Lectures, Tutorials, and Laboratories

Before listening to your lecture I was confused about this subject. Having listened to the lecture I am still confused, but on a higher level.

Enrico Fermi, Italian Nobel prize winner in physics (1901–54)

Key Concepts

- Listening to lectures
- Taking good lecture notes
- Scientific terminology and shorthand
- Getting the most from tutorials
- Why do we have laboratory sessions?

Introduction

We live in an information society. We are constantly bombarded with facts, figures, opinions, news, and views from a myriad of sources. We are expected to digest, translate, understand, and act upon this new information. Most is trivial and incidental and has no real impact on our daily lives. As a reflection of our technological society, university science-based courses are information-rich and dependent on content assimilation and interpretation. How do we cope with all this new information? Research with engineering students has shown that there are physical limitations that impact on the success of different modes of information transfer and that impact on what you do at university. This research found that we remember things at different rates, and that after 24 hours we remember about:

- 10% of what we read
- 26% of what we hear
- 30% of what we see
- 50% of what we see and hear
- 70% of what we say or discuss
- 90% of what we do.

With the above limitations, this chapter is an introduction to the three principal modes of information transfer in university science courses: lectures, tutorials, and practical classes. Electronic forms of learning are becoming increasingly important in most university courses and will be discussed separately in Chapter 5.

Lectures

Much of the information in a science-based course is presented as a lecture, which is an audiovisual presentation, particularly favoured when teaching large classes. The following applies whether you are either physically present in the lecture theatre or watching the lecture on an electronic device. About 25–30% of what you hear in the first half of a lecture will be forgotten by the second half. By the following day you will remember only about 30% of the lecture, and after three days you are likely to remember less than 10% of the content. So why do universities continue with the lecture format?

The main advantages of a lecture

1   Present information to many students in the most efficient way
2   Provide a structural framework for the main ideas
3   Explore and elaborate on areas of common difficulty
4   Forge links between prior knowledge and new knowledge
5   Summarise concepts, ideas, and abstractions
6   Stimulate and encourage further enquiry
7   Draw together information from a variety of sources.

A lecture is information presented in the format understood by the presenter. You, the listener, need not just absorb the information, but translate it into a format that makes sense to *you*. In order to make the most of the lecture you must develop a number of skills, perhaps the two most important of which are *listening* and *note taking*.

Listening

Firstly, some important background information about your sensory abilities:

1   We think in terms of mind pictures, which if translated into words would give us an equivalent *thinking* speed of thousands of words per minute.

2   The average speed of someone *talking* is about 125 words per minute.

3   The average rate at which we can *listen* to words and still comprehend their meaning is about 400 words per minute.

4   An average *writing* speed is about 35 to 40 words per minute.

So, in taking notes from lectures, you have a physical limitation on what you can *write* down, but considerable surplus capacity in what you can *think* about. You can reduce the physical gap between listening speed and writing speed by using abbreviations (you may wish to devise your own) and by being *selective* in what you record.

You can do very little about the lecturer's ability to deliver material effectively, but you can optimise your effectiveness in receiving the information. The key to this is to be an *active participant*. One simple way in which this can be done is to continually ask questions about the material being presented during the lecture. Simple questions such as:

*   Why is this information important?
*   How does that work?
*   Why does it work that way?
*   What is its relevance?
*   How is it connected to ...?
*   How can I use this information?

Lecturers often give verbal cues when they present a lecture, so be prepared for this by being tuned in to what the lecturer is saying. Some examples of cues could be:

Introducing the main ideas
*   'The crux of the theory is ...'
*   'The most important point I want to make is ...'
*   'The key concept for today's lecture is ...'

Giving examples of an idea already presented
*   'Let me illustrate this by ...'
*   'There are six examples where this theory is applicable ...'
*   'There is only one instance in which ...'

Using headings and subheadings for related information

- 'There are several factors that contribute to …'
- 'Firstly, there is the …'
- 'Next it was observed that …'

Summarising information

- 'In summary, …'
- 'Finally, …'
- 'So we can conclude that …'

You may never hear a lecturer use these exact phrases, but if you are tuned in to what to expect during the lecture then you will be well prepared for how the cues come.

TIPS FOR ATTENDING LECTURES

The following is a series of tips from university students on how to make the most of the lecture format.

Preparation

- Come prepared for the lecture by some pre-lecture reading.
- Review previous lecture notes so you know where the lecturer may be heading.
- Write down some questions from your reading that you hope the lecture will answer.

Tuning in

- Be physically and mentally prepared for the input of new information.
- Be an active participant; you are not there to have new knowledge 'inserted' into your brain.

Attention

- There are many possible distractions during the course of a lecture; be prepared and avoid them if possible.
- Do not let your mind wander too far thinking about things such as planning your weekend.

Questions

- Much understanding comes from asking the right questions.
- What is the main point of the whole lecture?

- How are the key ideas exemplified?
- How does the content relate to previous information?

Effective note taking

One of the key decisions you need to make for every lecture is whether to listen to the lecture, or attempt to write down what is said. Both listening and note taking are appropriate, depending on the lecturer and the type of lecture, but the following tips will assist you to take good lecture notes. Many, if not most, lectures these days have printed notes available that you can download from a website. Even if this is the case, the following information is still valid for working on the lecture notes to make them your own.

Mechanics

- Write notes; do not attempt to transcribe every word said.
- Write legibly, using headings, illustrations, and a code for the level of importance of the lecture material.
- Use page numbering and dates.
- Use abbreviations.
- Write on one side of the paper only and leave gaps between sections. Then you can fill in extra details or add other information from your readings. If you use loose pages, make sure you place them in an orderly fashion in a binder as soon as possible.
- Leave a wide (~4 cm) margin for 'tags' or headings, and for key comments. These will help you during revision.

Editing and reviewing

- Review your notes as soon as possible, preferably within 24 hours, to make sure they make sense.
- Improve the quality of your notes by working on them on a regular basis, for example over weekends.
- Add any necessary extra information as required from other sources, such as textbooks or reference books.
- Remember that the lectures are not the whole course but only a part of what you are required to know and understand. Use the lecture notes as the course synopsis.

Summaries

- At the end of each week write a summary of all the main points for each of the topics you study. Do this again at the end of each section of your notes.

- Use a different coloured pen for each summary.

- Use diagrams to summarise complex information and relationships.

Electronic lectures

Many lectures at universities today are available in electronic formats. The first and simplest of these are printed lecture notes, which you can download from the online site for your topic. Secondly, lectures may be recorded on DVD or VHS for students who are not able to attend because of timetable clashes or who are studying off-campus. These recorded lectures can be viewed through the library or department computer suite. Thirdly, there are lectures that are multimedia computer files. These can be viewed in real time at a distant location by video streaming, or they may be downloaded from an Internet site and played on your computer or a portable media device such as an iPod. All of these methods allow you hear and/or see the lecture as often as needed. This type of electronic lecture has many advantages over conventional lectures, but there is also a downside. As you *have* a copy of the lecture there is a tendency to think, 'I'll play it later, when I'm less busy', but it is easy to find other things to do that are more interesting that listening to a lecture.

Translations

One of the skills associated with lectures is the ability to translate complex or unfamiliar information into a format that you understand. For example, it is easier to learn several lines of a poem than a series of unrelated and unfamiliar words. Similarly, it is easier to learn information that has a visible pattern than that which has none. Look for the patterns and look also for the simple English meaning of complex information when writing notes for your own understanding. Remember that initially much of the information and the terminology used in lectures will be new to you. Do not let this be a deterrent. Understanding comes with familiarity. The following are a few examples of the type of language and terminology that may be used regularly in the sciences.

Consider the following statement:

Behaviour is a function of the interaction of an organism and the environment.

It can be expressed in plain language as:

How an organism behaves depends on its surroundings.

Much of science depends on the use of symbols, acronyms, abbreviations, or other forms of shorthand. For example, a mathematical statement used by chemists to describe the relationship between the variables that determine the behaviour of a gas is the Ideal Gas Law:

$$PV = nRT$$

This translates as:

The product of the pressure of a gas and its volume is equal to the product of the amount of gas, its temperature, and a numerical constant.

Other questions that may be relevant in understanding this equation are:

* What is an 'ideal gas'?
* What are the units in which the parameters P, V, n, and T are measured?
* What is the constant R and the units used to describe it?
* Does the equation always hold true?
* Why is it significant?

Scientific jargon is becoming a greater part of our lives through the popularity of television programs such as *CSI: Crime Scene Investigation*, daily news programs, and sci-fi movies. We are constantly exposed to terms such as DNA profiling, stem cells, blue-tooth, and gigabytes. So, not unexpectedly, learning in all fields of science requires you to become familiar with relevant terminology. Scientists are notorious for using shorthand to depict a range of ideas, from atoms and molecules, to systems of proportionality, chemical interactions, anatomical and physiological features, the names of diseases, and to physical conditions and dimensions. Many lecturers will use abbreviations and symbols without defining them all. Many scientific terms are derived from Latin and Greek words, from famous people's names; some are acronyms, and some are completely new words. A few examples will illustrate this.

Examples of scientific terms

Laser	Light Amplification by Stimulated Emission of Radiation
HTML	Hyper Text Markup Language
Robot	A word invented by Czech playwright K. Capek for a mechanical man
Fallopian tubes	Ducts which take the ova from ovaries to the uterus are named after the Renaissance Italian physician G. Fallopio who first described them.
DNA	DeoxyRiboNucleicAcid
MRI	Magnetic Resonance Imaging

English is a rich language, and there is often more than one way of expressing the same idea. This applies equally to the sciences. For example, the common drug paracetamol may also be referred to as acetaminophen, para-acetylamin-o-phenol, N-(4-hydroxyphenyl) acetamide, or by brand names such as Tylenol, Panadol, or Alvedon.

One of the key elements in attempting to write down lecture notes is to use abbreviations, symbols, or diagrams as a form of shorthand. Figures 4.1 and 4.2 give examples of a page of lecture notes taken during a nursing lecture and a chemistry lecture.

Figure 4.1   Lecture notes on key aspects of the *Circulatory System*

The circulatory system comprises three components:

1./ Blood.

About 8% of total body

Plasma — liquid part of blood (~55%) ← water / proteins / fats / sugars

Cells — Red (oxygenation) / Platelets (clotting) / White cells (defence) } haematocrit (40-50%)

2/ Heart

Muscle of 4 chambers.

Atria x2 / Ventricles x2 } separated by valves { tricuspid / mitral

1. aortic
2. pulmonary
3. auriculo-ventr.

Resting: 60-70/min

RA   LA
RV LV

Heart rhythme — cardiac cycle
heart rate/output
neural (vagus)   hormonal (adrenalin)

P   Q   S   R   T

3/ Blood vessels

Arteries (from heart)
Veins (to heart)
capillaries (gas exchange)

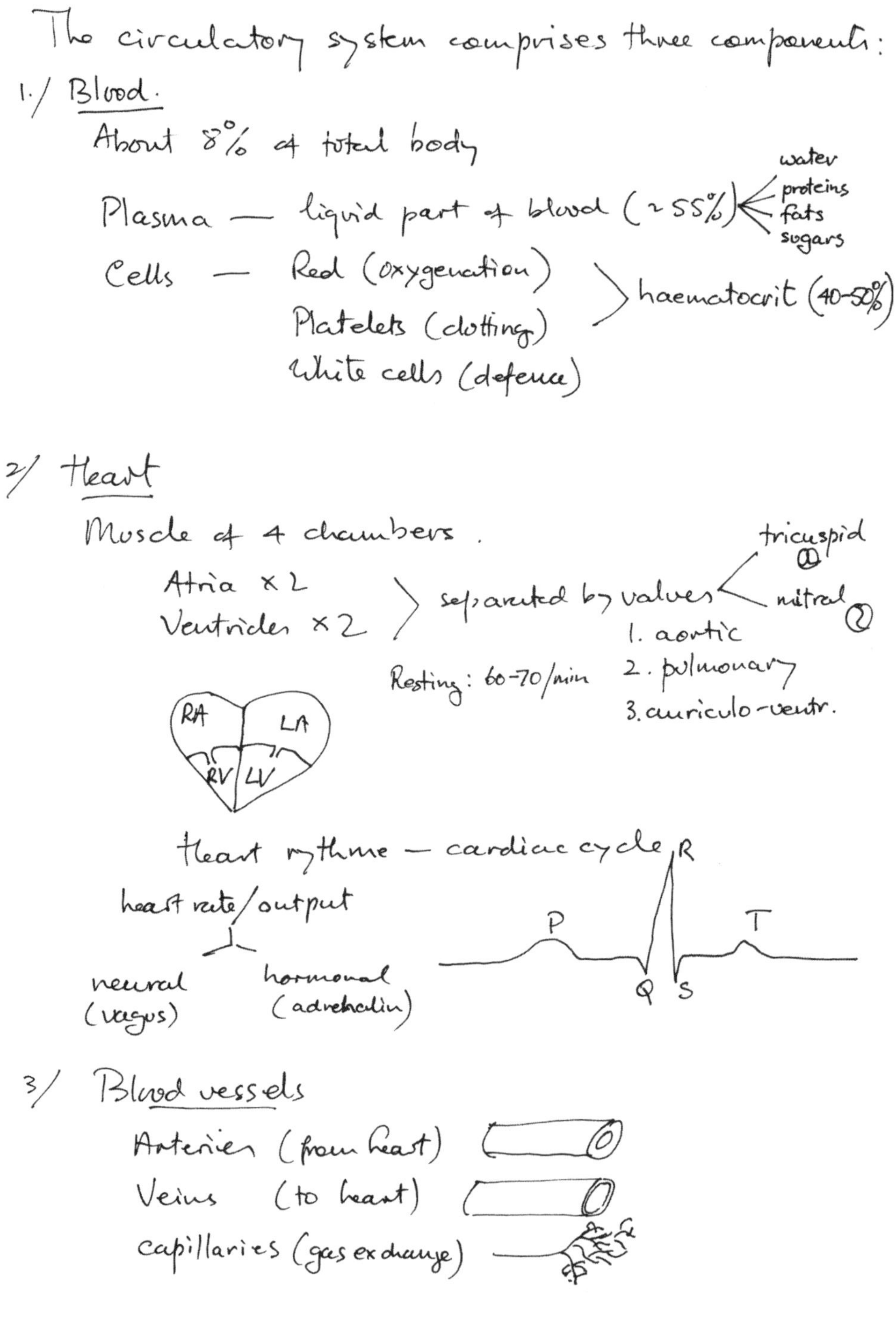

Figure 4.2   A schematic summary of a lecture on *Electrophilic Aromatic Substitution*

Electrophilic Aromatic Substitution:
  One of the key chemical reactions whereby parts of
an aromatic ring system can be substituted with
another element, or functional group.

Three key elements

1. Aromatic — occurs only in aromatic ring systems
               eg benzene, naphthalene.

2. Electrophile — is the reactive reagent that interacts
                  with the ring system.
               — may need to be generated, or activated
                  by catalysis in order to become active.
               eg $E \longrightarrow E^{\oplus} + e^{-}$ (catalysis, eg $AlCl_3$)
               — must be positively charged or have
                  strong positive polarity.

3. Substitution — some element/group in the aromatic
                  ring is replaced by E^+
                  $A\text{-}H + E^+ \longrightarrow ArE + H^+$

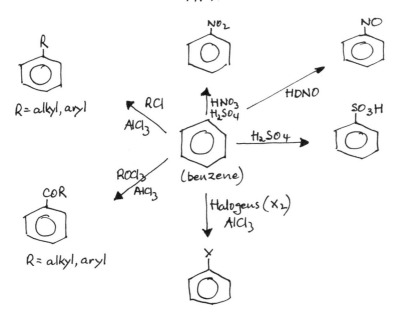

Laws of science

Science is about making logic of the universe by observation. It is not possible to measure everything in the universe, so scientists make approximations and propositions based on a limited number of observations. How good these approximations are depend on the number of observations, the accuracy of the observations, and the measurement techniques. There may be absolute truths in science, but equally there are many aspects of science that do not deal with absolutes. Several key terms in science that you will hear in your lectures or come across in your reading are defined below.

Some key terms in science

Hypothesis	A proposition based on reasoning and used as a starting point for investigation. For example, Einstein first hypothesised that time is relative to the observer.
Theory	An idea or premise used to explain observations that are derived from an hypothesis. For example, Darwin proposed the Theory of Evolution based on his observation of the changes in physical characteristics of birds and animals.
Law	A theory which carries with it a degree of certainty or general acceptance. For example, Newton proposed his laws of motion based on observations which he then verified with mathematical proofs.
Proof	Evidence in support of an accepted theory. In a sense, nothing in science can be proved beyond all doubt, as not all situations can be observed or measured, but it can still be accepted as truth until such time as disproved by more or better observations.

As more observations are made and more data collected and analysed there is a progression from ideas to accepted knowledge. For example, in the 'Law of Universal Gravitation' Newton first proposed the *hypothesis* that all bodies attract one another. His *theory* was that there is some universal force which accounts for this attraction. Newton tested his theory and developed a series of equations to describe his observations. Newton's theory has now been extensively tested, by scientists such as Galileo and even on the surface of the moon by astronaut Neil Armstrong. We now accept these observations to define a *law*. However, all laws are only as good as the observations that support them and many cannot be proven absolutely, as it may not be possible to conduct an infinite number of experiments to test each hypothesis under all conditions all of the time.

Tutorials

Tutorials in the sciences are somewhat different from those in disciplines such as the humanities or social sciences, where group discussion is an integral component of finding your 'own position' on an issue. The most common tutorial format in the sciences is usually structured around problem solving, where you are either given a series of problems to solve or previously set problems are discussed. This format tends to follow the tutor-centred approach where the tutor directs the approach and content covered.

However, you should not assume that all tutorials in the sciences follow this format, particularly with the growth of problem-based learning, where emphasis is placed on interactions within a group to collectively solve problems or discuss cases. Various forms of online tutorials are also popular in many science courses.

TIPS FOR TUTORIALS

In order to make the most of tutorials, keep in mind the following:

1. Attend

- Research with students in a first-year class showed that students who regularly attended tutorials improve their prospects of passing the course by 25%.
- Even if you learn only one new thing or have one aspect of a problem clarified, it is worthwhile participating.

2. Be Prepared

- Have at least one question written down to ask the tutor about in an area of the subject that you have difficulty understanding.
- Prepare for the tutorial by reading the lecture notes for the appropriate section.
- Have all the necessary materials with you.
- If there are set problems or exercises to be completed make sure you attempt them before the tutorial.

3. Participate

- The key to tutorials is participation. That means actively engaging and not sitting back passively hoping to absorb any useful information.
- Do not be put off by people who dominate tutorials; there's usually one in every group.

- Ask questions related to any of the topic material if there are things you do not fully understand. The tutor may give a different interpretation to the lecturer.
- Get to know your tutor and other members of the group.

Practical and laboratory classes

In essence all science is about observation, whether it involves a hands-on experiment in a geology laboratory, counting whales in the Antarctic, developing computer simulations for future climatic patterns, or learning how to use sophisticated electronic equipment for the diagnosis of disease. Most science-based courses involve some form of experimental or practical classes, which can serve a number of purposes, including:

- illustrating key theories
- becoming familiar with the language and instrumentation of experimental enquiry
- developing skills of experimental enquiry
- practising new techniques
- developing skills in observation and recording those observations
- collecting, collating, analysing, interpreting, and reporting data.

Examples of reasons we do practical work

Example 1

Joe is a second-year student in nursing who has excelled in classes in anatomy, physiology, and biochemistry and has just started weekly sessions on the hospital ward as an observer. Nursing, however, is about caring for patients, so in the first few clinical sessions Joe hopes to learn how to take a patient's blood pressure, clean and dress a surgical wound, insert an intravenous line, and continuously monitor patients' vital signs. Joe is expected to keep a daily journal of the practical experiences. Overall, what Joe will learn from these sessions are the basic clinical skills required of nurses everywhere: to observe, to record, and to act on those observations.

Example 2

Kim is a third-year student studying biological sciences with the intention of becoming an environmental microbiologist. Kim's next laboratory class involves the study of the growth habits of unicellular blue-green algae. Kim needs to set up the various growth conditions by changing environmental variables such as the amount of light, the ambient temperature, the clarity of the water, and the relative concentration of different nutrients. Kim then needs to make regular observations and record the findings. What Kim will expect to learn during this exercise is the methodology of biological experimentation, how to use instrumentation to measure and monitor microbial organisms, how to collect and analyse experimental data, how to interpret the findings of the experiment, and how to report those findings.

HINTS FOR PRACTICAL AND LABORATORY CLASSES

In order to make the most of the opportunities available to you as a student engaging in experimental and practical sciences, you should:

- be prepared for the session by reading ahead so that you come to the session knowing something about what to do, how to do it, and the reasons why you are doing it

- keep a notebook to record all your procedures, observations, and results

- record the date of everything you do

- consider the relevance of the experiment or practical session to the lecture course; keep in mind that not all laboratory or practical sessions are run after the lecture course has covered the same material

- find out how the report for the experiment or practical work needs to be presented for assessment

- keep all your marked and graded reports in an orderly fashion

- get to know other students in your practical classes and use each other as 'sounding boards' to evaluate your understanding

- if you get stuck on anything, ask for help.

Laboratory safety

Working in any laboratory can be potentially hazardous and dangerous. You may have to work with chemicals, micro-organisms, sharp instruments, or electronic measuring devices, all of which have a certain amount of risk associated with them. It is thus imperative that you follow ALL the required safety regulations associated with the use of a laboratory, including appropriate clothing (including footwear) and to behave in an appropriate manner. The laboratory handbook for each subject will have an extensive section on safety. It is important to follow all instructions at all times to ensure your and your classmates' safety.

Summary

This chapter has attempted to describe the main features of the three principal modes of learning in science-based courses. The key to making the most of each of the different learning situations is to be prepared to learn. This requires reading ahead, being organised, and striving for understanding. One of the key skills associated with any experimental or practical work is to write a report based on your observations. Writing a laboratory report is covered in Chapter 15.

5

E-learning and Research

The Greeks bequeathed to us one of the most beautiful words in our language, the word 'enthusiasm'—en theos—a god within. The grandeur of human actions is measured by the inspiration from which they spring.

Louis Pasteur, French microbiologist (1822–1895)

Key Concepts

- What is e-learning?
- Levels of e-learning
- The virtual library
- Using electronic databases
- The Internet

- The World Wide Web
- Online research using search engines
- Web-based learning

Introduction

'E-learning' (sometimes written as 'eLearning') aspires to be the future of education. Not surprisingly, the 'e' in e-learning stands for 'electronic', which seems a simple enough concept except that it is surrounded by much hype that mostly serves only to generate confusion. E-learning is a multifaceted and rapidly evolving mode of teaching, learning, and training. It is also something of a misnomer because *learning* can not be 'delivered'; it is a process that takes place within the learner. It is *information* and *instruction* that e-learning delivers electronically.

The genesis of e-learning was fuelled by the emergence of the Internet in the late 1980s. The advent of the World Wide Web (the Web), Web browsers, HTML, media players, and streaming audio and video began to provide a range of technologies to be explored and exploited in an educational context. The second wave of e-learning has resulted from technological advances such as high-bandwidth Internet access and computer network applications that allow ever-increasing amounts of data to be sent around the world at ever-increasing speeds, enabling the use of rich streaming media and high-speed communications so that live instructor-led training and mentoring over the Web make face-to-face interactions between learners and instructors (and learners and their peers) a viable proposition.

Five levels of e-learning

You will encounter some form of e-learning in your studies, either as a requirement of your study program or by choosing to supplement your studies by accessing the wealth of resources and information that e-learning technologies make available. Whatever the reason for your encounter with e-learning, it will take place at one or more of the following levels.

Information retrieval

While this is the most basic form of e-learning, it is also the most general and you should not underestimate its value and utility. There is a staggering amount of information now available online and it continues to grow exponentially. Learning at this level is apt to be self-directed, rather than following the guidance of an instructor, and you will likely use a Web *search engine* to find and retrieve information from the Internet. This may vary from largely conventional textbook style resources through to the latest research findings at the cutting edge of your field of study.

University libraries are still the hub of access to online material and they make it possible to search and locate texts, determine their availability, make loan requests, and search the catalogues of other libraries. Your library will also have a range of subject and journal databases that can be searched, including full-text resources.

TIP

Find out what your university's library has to offer: go on a library tour, learn to use the catalogues, find out what range of information is available to you, and how to access it. Don't be timid about asking the librarians for help in learning how to use their facilities. A relatively small investment of time and effort in the short term will prove excellent value in the long term, giving you the edge over other students who haven't developed these skills.

Knowledge databases

With knowledge databases we move from retrieving general information to accessing specific subject matter. If you are an Internet user, you will probably have encountered some forms of knowledge databases. These may vary from automated online help for a computer program, to frequently asked questions (FAQs), to an interactive database system that responds to key words or phrases.

Online support

This form of e-learning may either be self-directed or led (moderated) by an instructor. It is more interactive and reactive than a knowledge database, not least because the interaction occurs as communication between people rather than automated responses from a computer system. Online support takes a variety of forms such as forums (also known as bulletin boards or discussion groups), email, instant messaging, and chat rooms, all of which provide the opportunity to ask questions and obtain answers, guidance, and ideas from your instructor or other learners.

TIP

Find out what your topics have to offer by way of online support. Make sure you investigate all of these resources to see which ones suit your needs. It is also a good idea to do a Web search for sites at other universities as you are sure to find something useful. For example, a quick Web search will find online tutorials for statistics, how to use search engines, writing assignments, time management skills, and referencing systems.

Asynchronous instruction

Asynchronous means 'taking place at different times' as distinct from taking place in real time. In this mode of e-learning, you can access information electronically at times to suit your needs. For example, you may download lecture notes and read them when required, or you may access a complete lecture only minutes after it was presented. These lectures can be accessed from a PC using a media player program, or you may be able to download them as a pod cast.

In other forms of asynchronous instruction you may undertake specific computer-based activities and exercises designed to help you increase and test your understanding of course material. For example, using computer models can demonstrate physical laws or simulate laboratory experiments in a *virtual* environment which allows you to direct procedures, intervene, and respond in ways that mimic reality. Your input and general interaction with the software may be reported to an instructor so that your progress can be monitored. The simplest and most direct form of feedback to you and your instructor will consist of online tests and quizzes.

TIP

When you download lectures from the Web make sure you view the lecture as soon as possible because the information in this lecture will be assumed knowledge for the next lecture.

Similarly, when you download printed lecture notes, you must go through them and modify them so that they become *your* lecture notes. Having the lectures or lecture notes available online does not mean that you have learned the information until after you have worked on the material.

Synchronous instruction

This form of e-learning is like a 'virtual classroom'. Teaching and learning take place in real-time, led by a live instructor. Just as you need to attend lectures and tutorials at specified times, everyone logs in at a predetermined time to attend a virtual lecture or tutorial. The 'class' may take the form of a live broadcast or web-cast, some level of video-conferencing, perhaps through a virtual whiteboard with shared access or through specialised software running on your own computer or on a remote machine, which allows all class members to communicate in real time.

Now let us look at some of the specific examples of the electronic resources available.

The virtual library and electronic resources

All university libraries today consist of much more than just the physical buildings and the shelves of books. The library is a portal to all manner of electronic resources such as databases, e-books, e-journals, and the Internet. You can connect to these resources from a library terminal or off-campus from the computer at your home or work on a 24-hour basis. This is flexible delivery, as you can decide how, when, and where to access learning materials. So let us look at some of these electronic resources which the library has to offer some of which include:

Databases | Full text electronic resources | Lecture notes | Past exam papers | Internet resources | Library collections | Special collections | Subject guides |

By clicking on one of these headings we are directed to the link for that information source. For example, let's look at the link to databases.

Databases

The rapid rate at which knowledge is expanding requires new and fast ways of searching for and accessing new information. One of the most effective ways of doing so is through electronic databases. Databases are collections of information which may include reports, journal articles, books, and periodicals. Within science there are many databases; some of the better known and widely used include:

- Biological Abstracts
- Medline

- Computer Science Technology Report
- Proquest
- Current Contents
- PsychInfo
- Expanded Academic
- Web of Knowledge.

Each database has a number of options that can be selected to search for information on a chosen subject. Information is divided into fields such as:

Key words; Author(s); Book title; Book chapter, Journal name; Article title; Date

Let us say you are interested in looking for current information on robotics, and you decide to use the database called *ScienceSearch*. You link to the site using the LINK button to the appropriate URL and once in the site you enter key word = robotics. The database comes up with 2989 documents published over the previous five years. The database has listed the first 100 of these in ten pages (figure 5.1).

Figure 5.1   A database search result using *ScienceSearch*

ScienceSearch

topic = robotics
Products Searched = Web of Science; Current Contents. Timespan = Latest 5 years.

2989 document(s) matched the query. (100 shown)
Results Page 1 (Articles 1–10)
| 1 | 2 | 3 | 4 | 5 | 6 | 7 | 8 | 9 |

Click one of the product buttons beneath each result to view the full record in that product.

Let us now click on the first page of results, which shows the first ten documents. Figure 5.2 shows the first document listed.

Figure 5.2   Document details from search results

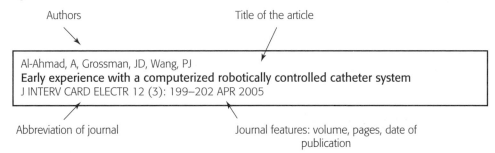

Authors   Title of the article

Al-Ahmad, A, Grossman, JD, Wang, PJ
Early experience with a computerized robotically controlled catheter system
J INTERV CARD ELECTR 12 (3): 199–202 APR 2005

Abbreviation of journal   Journal features: volume, pages, date of
                                                              publication

Some of the documents listed allow you to link directly to the electronic version of the article to read it online or download it to print later (figure 5.3).

Figure 5.3   Search results to a link

De Sapio, V, Warren, J, Khatib, O, et al.
Simulating the task-level control of human motion: a methodology and framework for implementation
VISUAL COMPUT, *eFirst*

Electronic link

Citation indexes

Most databases are designed to find what is currently in the literature. That is, they search for past works. There are other types of databases which look *forward* in time. This may sound a bit odd, but they are not looking for future events, but rather they can find works related to a previous work, but published since the first citation. Such a database is referred to as a *citation index*, two examples of which are the *Science Citation Index* (SCI) and *Google scholar*.

Let's look at an example by using the advanced search option of *Google scholar*. Say you are looking for articles that use information from an article on science learning by P. Zeegers, published between 2000 and 2005. In the 'Advanced Scholar Search' page, in the Author box we enter the author name and publication dates (figure 5.4).

Figure 5.4   Google scholar advanced search

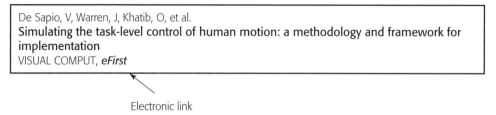

(Reproduced with permission from Google Inc.)

We then get a list of articles published by P. Zeegers between 2000 and 2005 (only the first few are shown in figure 5.5).

Figure 5.5   Google scholar search results

P Zeegers P Zeegers M Zeegers E Wagena P Knipschild F Rijsdijk	**Approaches to learning in science: A longitudinal study**—group of 3 » **P Zeegers**—British Journal of Educational Psychology, 2001— ingentaconnect.com Background. Longitudinal studies of students' approaches to learning in higher education can tell us much about the impact of the tertiary experience. ... Cited by 36—Related Articles—Web Search—BL Direct
	A Learning-to-learn Programme in a First-year Chemistry Class— group of 3 » **P Zeegers**—Higher Education Research and Development, 2001—Taylor & Francis ABSTRACT This article describes a national teaching project which set out to address the problem of high student attrition and failure in a rst-year ... Cited by 11—Related Articles—Web Search—BL Direct
	A Revision of the Biggs' Study Process Questionnaire (R-SPQ)— group of 3 » **P Zeegers**—Higher Education Research and Development, 2002—Taylor & Francis ABSTRACT The Biggs' Study Process Questionnaire (SPQ), an instrument for the evaluation of student learning in higher education, was revised over a ... Cited by 9—Related Articles—Web Search—BL Direct

(Reproduced with permission from Google Inc.)

The first article found is the one you may be looking for and it shows that the article was published in the *British Journal of Educational Psychology* in 2001 and has been cited by 36 other authors. If you then go to the 'Cited by 36' link, this will give a list of articles that cite this work (only the first two are shown in figure 5.6).

Figure 5.6   Google scholar search results citations

All Results A Duff K Lonka E Olkinuora J Mäkinen E Boyle	**Aspects and Prospects of Measuring Studying and Learning in Higher Education**—group of 4 » K Lonka, E Olkinuora, J Mäkinen—Educational Psychology Review, 2004— Springer Page 1. Educational Psychology Review, Vol. 16, No. 4, December 2004 (C 2004) Aspects and Prospects of Measuring Studying and Learning in Higher Education ... Cited by 10—Related Articles—Web Search
	Psychometric methods in accounting education: a review, some comments and implications for ...—group of 3 » A DUFF—Accounting Education, 2001—Taylor & Francis Page 1. Psychometric methods in accounting education: a review, some comments and implications for accounting education researchers ANGUS DUFF* ... Cited by 7—Related Articles—Web Search—BL Direct

(Reproduced with permission from Google Inc.)

The first example shows an article by Lonka, Olkinuora, and Makinen published in *Educational Psychology Review* in 2004 and which cited the 2001 work by Zeegers. So you can see these citation indexes are powerful and useful tools for research.

The Internet

The Internet is a fast and powerful way for people to communicate by using their computers, palm computers, or mobile phones. It is a system that links millions of sites worldwide. Information is sent electronically using modems and telecommunication lines, and more increasingly through wireless connections using microwaves. Electronic information which is downloaded can be displayed using languages such as HTML (Hyper Text Markup Language) or Javascript. The Internet, therefore, is another means by which information can be shared and accessed on a global scale, and this makes it an invaluable research tool. The Internet is an almost limitless source of free information, but there are many sites where the user pays.

The World Wide Web

The World Wide Web was first developed in 1990 at the European Organization for Nuclear Research (CERN) located near Geneva in Switzerland. It is a system for displaying information electronically. This information can be displayed in a number of forms, including text, graphics, sound, or video. Information is organised on Web pages which can be written by individuals, companies, or organisations using web-authoring programs.

The WWW is a bit like being inside a huge virtual city, with millions of buildings each of which may contain a variety of information, which may be pictures, sound, video clips, or computer programs. Each building has doors or links to other rooms that are usually highlighted or a different colour.

Accessing the WWW requires the use of programs called Web browsers, which connect your computer through a network or a telephone line to an *Internet Service Provider* (ISP) that displays information retrieved from computers on the Internet. The two most commonly used Web browsers are *Netscape Navigator* and *Microsoft Internet Explorer.*

Some of the different types of information sources could include:

- Libraries (such as the British Library, the Library of Congress)
- Scientific periodicals (*Nature, Science*)
- Government agencies and archives (Australian Bureau of Statistics, NASA)
- Commercial sites (Carsguide, Microsoft)
- Universities (University College London, University of Hong Kong)
- Newspapers and other periodicals (*The Times, Washington Post*)

Using search engines for research

To find information on the WWW you need to use *search engines*. These are like indexes that search particular parts of the WWW for the key words that you type in and then present you with websites to which you can connect. It is important to become familiar with a range of search engines because they use different types of databases and so may come up with different results for the topic that you are looking for.

There are a number of search engines, the most popular of which are listed in figure 5.7.

Figure 5.7   Search engines

Commercial search engine	
Ask	http://www.ask.com
Google	http://www.google.com.au
Surfwax	http://www.surfwax.com
Yahoo	http://www.yahoo.com
Academic search engines	
Google scholar	http://scholar.google.com
Windows live academic	http://academic.live.com/

Different search engines can be useful for different purposes, so if you want to find out more, a good online tutorial is available at:

http://www.lib.berkeley.edu/TeachingLib/Guides/Internet/FindInfo.html

It is also worth mentioning that although most search engines are not case sensitive, that is, they do not distinguish between upper and lower case, there are still a few search engines that are. It is therefore safest to type all search terms in lower case (no capitals) because this will always retrieve any upper case terms as well.

How to sort information

There is a vast amount of information on the WWW. Although much of it is extremely valuable and interesting, there is much that is trivial, blatantly commercial, worthless, and incorrect. It is therefore vital that you *critically evaluate* any information found by assessing its accuracy, reliability, and honesty. Who created this site? Whose interests does it serve? Is the site a commercial one that is really just trying to sell you something? Has the information on the site been validated in any way? How can you be sure that the information is accurate?

In searching the WWW, you are searching through information that has been specified by the search engine, and not necessarily by you, the searcher. The success of your searches therefore depends on three factors:

1   your ability to create exact matches between the terms you search for and the terms used in the information you wish to find

2   the contents and validity of the database chosen

3   the features available for searching the database.

This means that you may have to think of variants, synonyms, and related themes to get what you want. Otherwise, if you only search for commonly used words, you may get irrelevant documents containing your words but not your subject.

Improving your search results

All search engines search through Web pages looking for key words of phrases nominated. Some search engines allow you to use *Boolean* logic which may help you to narrow down your search. This means using terms such as 'AND', 'OR', or 'AND NOT' before and after your search terms. These Boolean terms are also sometimes represented by the symbols '+' and '–' instead of 'and' and 'not'.

Google is probably the most popular of the search engines and can find vast amounts of information. However, for the academic work expected at university, most of the sites found using Google may only be marginally useful. To limit your searches to validated and published works of academic quality, Google has developed the *Google scholar* search engine. So let's compare the search results using these two search engines.

A search example

Let's say that you have been asked to write a 2000 word essay on modern methods of bowel cancer screening. You decide to use *Google* to search for relevant information. To start with you could type in each of the three terms: bowel, cancer, and screening. *Google* comes up with 1 160 000 sites which contain these key words. Many of these sites are health agencies, government departments, self-help groups, and commercial sites. The nature of the information is likely to be very general, and may not be very helpful. One way to narrow down the search is to type in the key words in a string as 'bowel cancer screening'. Using this string, *Google* now finds only 68 000 sites, but the majority of these are still of the same type as before, and are not suitable to use in an academic essay.

Now you repeat this process using *Google scholar*, which will only search through peer reviewed or evaluated material found in published works. Putting in the same three key words, *Google scholar* now finds 23 500 articles. You then repeat the search using the three-word string and this time you find 178 published works going back to 1995. You are mainly interested in the latest developments in screening techniques, so in the *advanced* section of *Google scholar*, you limit the search to works after 2002. This time you find 87 articles published since 2003. So you can see that it important to know exactly what it is you are looking for and use the right logic to narrow down your search.

Google	key words	1 600 000	sites of all types
Google	string	68 000	sites of all types
Google scholar	key words	23 500	published works
Google scholar	string	178	published works
Google scholar	string (date limited)	87	published works

Be aware that when using most of the common search engines for information on a wide range of subjects one of the sites that invariably comes up is the online encyclopedia *Wikipedia*. This site can be a very valuable source of general knowledge information but, as there is no strict validation system for any of the information posted on this site, there is no guarantee that the information is accurate or correct.

Web-based learning

There are a number of web-based learning programs that are widely used at universities around the world. Some of the better known are *Scholar360*, *Blackboard*, and *WebCT*. All of these programs allow online interaction between the learner and the instructor. These are web-based course management systems which provide an integrated range of services including:

* presentation of course content (lecture notes, examples, ilectures)
* learning support (quizzes, tests, online support)
* interaction of participants (chat rooms, email)
* electronic submission of assignments (these may use anti-plagiarism software)
* course management (class lists, marks, grades).

Let's look at what is available using these web-based learning platforms. Firstly, the information is organised around a Web homepage. Once you gain access (login) you will be taken to your personal home page and you will be able to access all your topics by clicking on the name of the course. This homepage is the entry point for the web-based learning and is the first page that you see after logging in. Login to the Web pages for a particular topic is only available when you are recognised as a student enrolled in that topic. Resources available at this level include, among other things, a text message box, links to course content such as lecture notes, quizzes, and assignments, and links to course tools. Each topic will have icons that link to tools and materials.

Learning tools

Web-based programs provide a range of tools to support the educational process. Some have a 'more' link for further information about the tool. Examples of the most common facilities available through your homepage include:

- **Discussions**

  This tool allows communication among students, topic coordinators, and tutors. Messages can be arranged under topics related to their contents. Initially, only unread articles are displayed. Messages can be searched for content, sender, date of sending, and more. Messages can contain links to external Web pages.

- **Mail**

  Mail allows electronic messages to be transferred among course participants. The mail tool does *not* integrate with other Internet mail. While messages can be redirected to an Internet email address, Internet email cannot be redirected to mail.

- **Assignments**

  The assignments tool displays information about assignments and can have additional resources attached. The tool allows students to submit assignments electronically for marking.

- **Quizzes**

  A quiz is a test for which grades are assigned and which is created and administered online. A quiz can be marked and graded by the program. Once a quiz is started, a clock counts down the time left to complete the quiz. Once completed and marked, grades and comments are made available.

- **Self-tests**

  Self-tests are added to a page of content and are used to test your knowledge of the subject. Responses are automatically marked as correct or incorrect. Extra information may be provided for each answer, which can be viewed when an answer is marked.

- **Chat rooms**

  Chat rooms provide real-time communication among course participants. Be aware though that your conversations can be accessed by the topic coordinator. There may be other chat rooms for you to communicate with other students not in your topic.

Net etiquette

Finally, a brief note about using appropriate etiquette when on the Internet. When you communicate with others electronically, mostly all you see is a computer screen. You don't always have the opportunity to see expressions, gestures, and tone of voice, so it may be easy to misinterpret your correspondent's meaning, and they yours. The person on the other end is someone like yourself, so you should show them and expect in return the same level of moral, ethical, and personal behaviour in cyberspace as you would expect if you were talking to the person face-to-face.

Summary

In essence, e-learning is electronically delivered instruction that can take place any time and anywhere, with course materials and up-to-date information able to reach large numbers of students in diverse locations. It can include text, video, audio, animation, and virtual environments that, in combination, can provide a rich and engaging encounter with learning materials and has the potential to surpass the classroom experience. Remember, however, that in order to make the most of this information you need to develop your critical and analytical skills.

6

Assessments

In examinations the foolish ask questions that the wise cannot answer.

Oscar Wilde, British novelist, poet, and playwright (1854–1900)

Key Concepts

- Grading systems
- Assessment types
- Why we do exams
- Tackling exams
- Memory and learning
- Exam tips
- Multiple-choice exams
- Maintaining a positive attitude

Introduction

Most of us aspire to getting a driver's licence and driving our own car. To do so we have to learn all the road rules and then pass a test in which we are expected to demonstrate that we can safely drive a car on the road. Part of the process of the acquisition of any new skills and knowledge is the need to evaluate whether or not this acquisition has been successful. This is done by means of assessment, of which there are fundamentally two different types. Firstly, there is assessment *for* learning, referred to as formative evaluation, and which occurs during the learning process. The methods of formative evaluation employed at university will vary from one course to another. The second form of assessment is assessment *of* learning, often referred to as summative evaluation, which occurs at the conclusion of the learning process. The most common form of summative evaluation at university is the examination.

Grading systems

The grade for a topic is the sum of all the individual assessment tasks for the topic. You should be given a printed copy or have electronic access to this information for every topic in which you are enrolled. As an example, let's consider the assessment requirements for Environmental Biology 2 shown below:

Assessment for Environmental Biology 2, 2007

Weekly online quizzes (best 10 are counted)	20%
Laboratory reports (3 written reports required)	20%
Oral or poster presentation (10 minutes)	10%
Final examination (3 hours)	50%
TOTAL	100%

At the end of each study topic you will be awarded a grade, which at most universities takes the form of a letter or word grade. An example of a possible grading system is shown in figure 6.1. Be aware that this example may not be the system used at your university.

Figure 6.1   Example of a grading system

Final topic score	Letter grade	Word grade	Abbreviation
85% and over	A	High distinction	HD
75–84%	B	Distinction	D
65–74%	C	Credit	CR
50–64%	D	Pass	P
~50%	NGP	Non graded pass	NGP
Less than 50%	F	Fail	F

Some universities award a grade called a *terminating pass*. This means that you may be awarded the points or units for this topic as a passing grade, but that you may not use this topic if it is required as a prerequisite for further study in that area of study. For example, you may be awarded a terminating pass (~50%) for the first-year topic Chemistry 1001, but you are not able to continue studies in second-year chemistry topics.

If you withdraw from a topic before the predetermined date, you will be awarded a withdrawn (W) for which there is no penalty, but if you withdraw from the topic after this set date, it will count as a withdrawn-fail grade (WF). The cumulative total for the grades you receive for all the topics you study in any one semester, and for the whole academic year, is called the Grade Point Average (GPA). The most common form of GPA is a number system from 0 (lowest mark) to 7 (highest mark).

Continual assessment

Research data indicates that students often do better in their studies when they have regular assessments covering small amounts of course content. The types of continual assessment tasks in different courses are likely to be quite variable, so it would be an impossible to cover all these methods here. Some of the more common forms include written assignments, quizzes, multiple-choice tests, essays, online assessments, practical exercises, and laboratory reports. All continuous assessments are used as part of the total assessment for your topics, but they also serve another important function, and that is to allow you to *assess your own level of knowledge and understanding* of topic content integral to your course.

If you are studying in a course that does not have a continuous assessment format it is worthwhile setting your own study problems and giving yourself weekly quizzes to help prepare you for major tests. Better still, get together with some of your classmates and set up a peer support group where each of you sets a series of problems. For example, in say a group of three, each student has the task of developing five questions or problems based on the previous week's course content. The students setting the questions must also provide appropriate answers. Each student attempts the other student's questions, and at some time a few days later you get together to discuss the questions and the answers.

The one type of assessment that all of you will encounter is the written examination, so the rest of this chapter is devoted to this topic.

Exams

Most of you know what exams are as you have probably done many of them in your previous studies. Many students dread exams, but some look forward to them as an intellectual challenge. To put exams into perspective, consider that you are flying to Fiji for a holiday and the plane is about to land at Nadi airport. At some stage you are likely to think to yourself,

> *I wonder if the pilots passed the landing exam. Did they only get 50% of it right? I hope the landing exam he did was very demanding!*

So you see it is important for your own peace of mind to know that the pilot has successfully acquired a high level of the specific skills required for that profession, that the pilot has been thoroughly assessed in those skills, and, finally, that the pilot did more than just pass exams through rote learning.

Exam types

The type of examinations you come across will vary from one course to another, but the most common exam types are:

1   the 'typical' exam, where no resources are available to you except something to write with, and perhaps a calculator

2   the 'open-book' exam, where you are allowed to take into the exam room lecture notes and maybe textbooks

3   the oral exam, which tends to be more popular in postgraduate study where students are asked to defend something they have written

4   the online exam, where you answer questions at the computer from a database of questions, and where you may get immediate feedback.

A good university exam should be a test of your understanding of the scientific skills, knowledge, and principles underpinning your area of study, as well as your ability to apply that understanding to new situations. Good exams are difficult to write and many lecturers spend a great deal of time constructing what they believe is a fair and reasonable evaluation of your knowledge and understanding in their topic.

What is an exam designed to do?

An exam can be a test of your ability to:

1   understand the basic skills, knowledge, and principles underpinning your chosen topics

2   recall important facts, structures, figures, symbols, reactions, formulae, effects, and relationships

3   apply your knowledge and skills to make a deduction, to construct a logical argument, or to propose a solution for a particular situation or problem

4   interpret data critically, to appraise it and to determine its value; the type of data could be graphical, tabular, flow charts, equations, diagrams, blood screen analysis, or CT-scans just to name a few types

5   extrapolate your current level of understanding and knowledge and apply it to a new or hypothetical situation

6   work successfully under pressure.

Because of the multitude of things that an exam can be designed to assess, there are a number of formats exam questions can take in order to achieve these aims, some of the more popular are:

1   *Multiple-choice questions*, which are particularly favoured for large first-year classes. These questions usually provide you with a choice of several possible answers, only one of which is correct (or incorrect).

2   *Short-answer questions*, which ask you to recall fundamental information such as terms, relationships, characteristics, or terminology. They tend to contain verbs such as define, describe, list, recognise, or demonstrate.

3   *Problem-solving questions*, where specific information or data is provided and you are expected to use this information to arrive at a known solution. These questions tend to contain verbs such as solve, calculate, find, or determine.

4   *Short-answer questions*, where you may have to explain a theory, an observation, a chemical reaction, a biological effect, or a physical phenomenon. These questions tend to contain verbs such as compare, attribute, differentiate, or summarise.

5   *Essay-type questions*, where you may be asked to construct a logical argument or to give detailed information about the factors which may impact on a given scenario and how they may be interrelated. These questions tend to contain verbs such as analyse, predict, judge, construct, or design.

Tackling exams

For any assessment task there are three broad principles which, if applied diligently, will give you the best opportunity of success. These are preparation, technique, and confidence. How you go about acquiring each of these skills, however, is largely a matter of trial and error, but the following information is designed to give you some suggestions, which will allow you to build on what you already know and thus help you to develop the right approach for your needs.

Preparation

The first thing you need to establish is your approach to studying. You need to determine the right time of day to study, where, how long, with whom, and what method to use. At this point it may be worthwhile to refer back to the different methods of study discussed in Chapter 3. Ultimately, preparation is about effectively utilising all the resources available to you and working at your optimum level of efficiency.

Resources

Many resources are available to assist in your preparation for assessments, some of which include:
1   topic book, syllabus, lecture notes, tests, and problem sheets
2   teaching staff, including lecturers, tutors, and demonstrators
3   laboratory notebooks, manuals, and reports
4   online materials such as self tests and trial exams
5   past exam papers (usually available in the library)
6   other people, including friends, study groups, and students.

If there is something you are not clear about, ask someone for clarification. Don't give up too easily if help is not always immediately at hand, and remember that when you share information in the correct manner with others you will gain as much as you give.

The revision plan

No matter how simple or how complex your plan for revision, make sure you have a written plan. Allocate your time so that all topics are covered and you don't neglect the hard or boring parts. Set aside some spaces for 'free study' to allow yourself to go over difficult work or to catch up on subject areas for which you may not have allotted enough time. Do not start with the most difficult topics. Start with easy ones, or review some previous work, as this will increase your confidence. It is easier to tackle the more difficult areas when your confidence is high. So the things you need to do are:

- Draw up a revision timetable well in advance, and try to stick to it.
- Select, categorise, and summarise all relevant information in a way that suits your needs, and so that it makes sense to you.
- Set specific tasks with deadlines.
- Spread your study periods out over the whole week, all *seven* days.
- Be flexible, adaptable, and, above all, diligent.
- Have regular study breaks each day for recreational time and 'downtime' to give your mind a break and prevent you from becoming stale or bored.
- Allocate more time for the topics you find difficult.

Techniques

Strive for understanding

The essence of all learning is to understand. This involves more than just 'knowing' the topic content but involves your comprehension of the interrelationship between the topic content and how to utilise it. Perhaps the two most important words to test your understanding of topic content and the underlying fundamental principles are *how* and *why*. You are more likely to remember specific details if you can see how and why they fit together, rather than if you are just trying to recall vast amounts of (apparently) unconnected facts. One method to assist your understanding of course content is to relate what you know to the course aims and objectives. As an example, consider the following two passages:

> *Disruptions affecting the diencephalons, midbrain, pons, and medulla usually cause a predictable pattern of change in the level of consciousness. This change is indicative of bilateral hemisphere damage with the danger of tentorial herniation. There is some combative movement as well as movement in response to pain.*

> *The head bone is connected to the neck bone, the neck bone is connected to the shoulder bone, the shoulder bone is connected to the back bone, now hear the word of the Lord. Them bones, them bones, them dry bones. Them bones, them bones, them dry bones, now hear the word of the Lord.*

Which paragraph would be easier to learn and understand? Why?

Be an active learner

Just reading textbooks or lecture notes and underlining key words, phrases, or relationships is helpful but not sufficient. You must *work* on your notes, textbooks, and other information sources. In order to do this you need to organise, summarise, and condense your notes as you go. Each time you work on your notes you will gain a firmer grasp on the material and understand and remember a little more. You find common themes and links between information from different sections.

Remembering

One of the fundamental skills of learning, and one often undervalued, particularly in non-science courses, is the skill of memorisation. Learning is a continuous and sometimes slow process requiring regular input over time. Once basic principles are assimilated and understood it is easier to remember the details and then apply them to new situations. Unfortunately, we tend to forget what we do not use regularly, so the best way to avoid this is continual reinforcement. It is important here that we make a distinction between memorising and rote learning. Perhaps some examples will illustrate this.

Examples of memorising

Example 1

Two people are given a copy of Shakespeare's play *Richard III*. The first person is your local butcher, Fred Mutton, and the second is the actor, Tim Crooz. Each is told to memorise the role of Richard and asked to come back in a week and be able to recite it from memory. Both come back in a week and recite the words verbatim, but how they use the words may be very different. The role of the actor is to use the words as a tool to give the play meaning, so he has memorised it with a specific purpose in mind. To Fred the butcher, it may be purely an exercise to see if he can remember all the words. If, on the other hand, both people were given a list of all the different cuts of meat and recipes to use them to cook exotic meat dishes, then the result may be different. It's all about relevance.

Example 2

Radiologist Dr Clare Thinkin has taken a series of CT scans of the head of a patient with persistent and severe headaches of unknown origin. When describing the CT-scan of the patient's head to the neurologist, Dr Cranium, she uses the anatomical and pathological terminology she has memorised during her radiology degree. Because of the context of the terminology, and because both parties are familiar with it, she can explain the meaning of the scan and the importance of the findings to the well-being of the patient. Dr Thinkin has memorised the terminology for a purpose.

There are many methods you can use to help you to recall information, some of which are listed below.

- Rewrite lecture notes and make summaries, then summarise the summaries.
- Redo tutorial and assignment questions.

- Do trial exams, past exams, or test questions set by friends.

- Transpose information into a format that makes sense to you, so that it becomes 'your version' and thus you are able to use it appropriately.

- Memorisation of some information simply requires rote learning. This could include all those 'must know' equations, rules, names of structures, relationships, and formulae. One way to help you to remember this information is to make filing cards for this information so that you can read them anywhere, any time (see Chapter 3).

- Other methods to improve your recall ability is to read out loud, chant, or sing information, devise your own mnemonics, and use abbreviations or acronyms.

- Make visual aids. Research has shown that visual memory is often better than word memory. In order to make the most of this, make your own visual aids in the form of diagrams, flow charts, tables, or mind maps.

- Hang up posters containing relevant information in places you visit frequently or where you spend a lot of time, for example, on the fridge, above your desk, or above your bed.

Self-confidence

The third key to success in assessments, and particularly for exams, is *confidence*. Confidence is a strange and fragile creature that can take years to develop but only minutes to destroy. Ultimately, confidence comes from within and in order to give yourself the most self-confidence there are a number of things you can do.

- Maintain a positive attitude, even towards exams. An exam is about finding out what you know, what you understand, and whether you can apply your skills and knowledge. An examiner could easily set an exam that almost nobody could pass, but what would be the point? So stay positive.

- By just being at university you have already demonstrated to others and to yourself that you have the intellectual capacity for university level study.

- By continuing to become more and more familiar with course content through repetition, summaries, self-examination, and course feedback your self-confidence will grow.

- If you have regularly attended and actively engaged your brain in the majority of lectures, tutorials, and laboratory sessions, then you will have covered all the course content.

- If you have made an honest effort in all your assignments, tests, and tutorial problems there is no reason you should not understand the nature of the materials covered.

- Studying at university is just one means to an end, whatever your long-term ambitions may be. Failure in a single topic or even a whole course is not the end of the world. Learn from your failures.

About the exam
Format

Look at past exam papers to determine the sort of questions previously asked to see if they are multiple choice, short answer, problems, recall, interpretation of data, or a combination of these. Will the format be the same this year? If you have not been told, then ask!

Length

Most exams at university are two to three hours, but find out!

Value

What proportion is the exam for the total topic evaluation? How well are you doing in the other assessment components for the topic? You should be able to determine what score you need to achieve in the exam in order to attain your desired level of success in the course as a whole. Set yourself a realistic target result based on all the information at hand.

TIPS FOR THE EXAM ROOM

There are a number of simple things you can and should do once you are actually in the exam room with the exam paper in front of you.

- It may sound obvious, but check the paper to make sure it is the right exam for your topic.

- Spend a few minutes reading the requirements of the exam. Look at how many questions there are and determine their relative value. Look also to see if all questions are compulsory, or whether you have a choice.

- Make a rough schedule of your time and allocate the appropriate amount of time to each section or question. For example, if the exam is three hours and there are five sections, and each is of equal value, then allocate 30–32 minutes to each section. This will allow you about 20 minutes at the end to go over each section again and make changes.

- Jot down essential information next to each question as you read it; this will help you answer the questions when you write your answers in the answer book.

- Do the easiest questions first and the most difficult questions last, as this will put you in a positive frame of mind and increase your confidence.

- Make sure you answer the question asked of you, and do not be tempted to write down anything or everything you know that may in some small way be related to the question.

- When solving problems that require you to manipulate data, do not be in too much of a hurry to just use the numbers. Think about the problem and the nature of the answer required. For example, say you are required to calculate a drug dosage involving concentrations and volumes. Make a 'guesstimate' by determining the approximate order of magnitude of the right answer. Is the answer likely to be in millilitres, litres, or kilolitres?

- Tick off questions as you complete them to ensure you have completed all that is required. Keep track of your time in the same way.

- A blank answer will score you no marks, so attempt to answer all questions. If you run out of time and have not finished, write down key words, equations, and relationships you would have used and elaborated on, to show the examiner that you know the essential information. Examiners want to pass students so you may get some marks for at least attempting an answer.

- Above all else, do not panic, for in so doing you become your own worst enemy.

The multiple-choice exam

For many students the multiple-choice exam seems to evoke a sense of fear, dread, or loathing. There is no single trick to doing this type of exam question but there are some fundamental principles that you need to bear in mind.

- Is there a penalty for incorrect answers? If the answer is *no*, then feel free to make a guess if you are not sure of the answer. If the answer is *yes*, then only answer the question if you are reasonably confident that your answer is correct.

- Read the question and all the answers thoroughly. Do not jump to conclusions.

- Make sure you fully understand the nature of the question and the optional answers. Sometimes you are asked to select the answer that is *most* correct and at other times the answer that is *not* correct.

- Science can at times deal in absolutes and use terms such as all, none, always, or never. For example, consider the question, '*Gaseous hydrogen chloride completely occupies the volume of a closed one-litre container in which it is held*'. Some of the possible answers may imply that certain factors such as temperature or pressure may restrict the volume of the gas, but a gas will *always* occupy the volume to which it is restricted independent of other variables.

- Some examiners like to include answers that are bogus, silly, or clearly wrong. These answers are sometimes added as a bit of comic relief, or because the examiner ran out of better alternatives.

- If out of a choice of four possible answers, two seem to be most likely, try to choose the one that most closely fits the data. In order to decide between the two most likely answers you may have to reread the question and each answer again and again. Sometimes this process of repetition helps to clarify the distinction between the two answers.

- Many multiple-choice questions rely on your understanding of the essential terminology for the topic, so the more familiar you are with this terminology the more easily you will be able to answer the questions.

- Each question and its answer are *independent* of all previous questions, in that there is no pattern to the selection of the right answers. So if you thought that the correct answer to the first five questions was choice (b), then there is no reason why the correct answer to the next question cannot also be (b).

Summary

Assessment is an integral component of the learning process and one that should not be feared. A combination of the three key skills of preparation, technique, and confidence should guide you successfully through most assessment tasks, particularly if you are prepared to:

- commit yourself fully to all your learning tasks
- work to a varied, flexible study plan but, above all, have a plan
- strive for understanding and use memorisation as a tool for learning key information
- rework your lecture notes regularly, make summaries, and create 'pictures' as memory prompts
- keep a positive attitude, be efficient, and think things through
- seek help when the need arises from teaching staff, friends, or other students.

PART 3

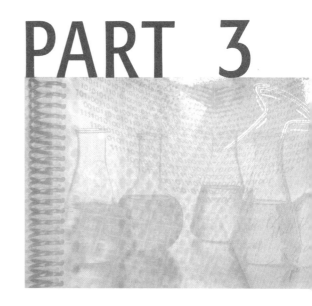

Critical
Evaluation

7

Reading Scientific Literature

There are two motives for reading a book; one, that you enjoy it; the other, that you can boast about it.

Bertrand Russell, British philosopher (1872–1970)

Key Concepts

- What is the scientific literature?
- Reading concerns
- Hints to improve your reading skills
- Analysing a scientific article

Introduction

One defining feature of all technological civilisations is the development of the written word. Writing in its many and varied forms has been used for thousands of years to record important events, to communicate thoughts and ideas, for bookkeeping, and for telling stories. The earliest forms of writing were pictographs using coloured earth and fire ash on the walls of caves. These told the story of a successful hunt or other important events in the lives of those who recorded them. More sophisticated forms of writing were developed about 6000 years ago in the Middle East and China. These were a portable form of writing that used clay tablets, animal hides, metal scrolls, and, later still, pen and ink and various forms of paper. With more sophisticated forms of writing, a catalogue system was needed to collect, collate, and store these written records and make them accessible to future generations. Consequently we have the birth of the concept of a library.

Ultimately, all scientific research is about discovering more about ourselves and the universe in which we live. Each step we take along the road of discovery needs to be recorded. This becomes a recorded scientific history of where we have been and what we have seen along the way. Scientific knowledge is not static but is ever changing, so this recorded history not only provides us with a glimpse of the past but also a platform for the future. Our recorded history provides us with a guide as to where to go next. This recorded scientific history is referred to as the *scientific literature*, and this chapter will guide you through this scientific record and show you ways in which you can read and understand it.

The scientific literature

The scientific literature is often divided into four major components according to the age of the information.

Textbooks, reference books, and handbooks

The oldest scientific information is that found in textbooks, reference books, and volumes of technical data. This information is regarded as accepted knowledge.

Examples

Textbook of Toxicology

Dictionary of Programming Languages

Handbook of Chemistry and Physics

Reviews, monographs, and annuals

Information and knowledge from recent research can be found in specialised volumes called annual reviews, synopses, or monographs, where research is summarised, reviewed, and critiqued.

Examples

Annals of the Academy of Medicine

Review of Biochemistry

Philosophical Review

Journal articles

The results of current research work are published in periodical form, which may be journals, communications, or letters. These are published monthly, fortnightly, and sometimes weekly, either in hard copy or as electronic journals. Before an article is published in a journal it goes through a process of *peer review*. In this process the authors send their article to the journal editor who in turn sends it out to be reviewed by two or three independent reviewers, who are scientists knowledgeable in that area of study. It is the task of the reviewers to validate the article in terms of its methodology, logical arguments, data analysis, and structure. Only if the reviewers consider the article as adding significantly to our scientific knowledge is the article accepted for publication. Many more articles are rejected than are accepted.

Examples

Journal of the American Chemical Association

Communications in Behavioural Biology

Physics Letters

Scientific abstracts, citations, and electronic databases

In order to keep track of who is doing what research, a number of databases are available. Most databases are now in electronic form and can be accessed through library websites, either in alphabetical order or grouped by subject headings. Some databases contain the synopses of research findings, some give a list of abstracts and the publication information (author, date, journal, and publisher), while some others link to complete articles. In general, databases allow you to search for relevant information, cross-reference work, or simply negotiate your way through an ocean of information.

Examples

Medline

Current Contents

Proquest 5000

Expanded Academic

Part of any writing project is to learn how to effectively retrieve information from this huge volume of work. Each type of literature source has its place in the context of a written assignment provided that you keep in mind the type of material and its age. For most literature, also bear in mind that what you read can be biased to highlight the author's point of view or interpretation of research data. You, the reader, need to be convinced that the arguments presented by the author are persuasive and supported by the data. Thus, one of the most important skills you need to develop as a tertiary student of the sciences is the skill of reading critically.

Reading concerns

Many students, particularly those in the early years of higher education, have some concerns about reading highly specialised literature. Some of the common concerns expressed by students when they first start reading scientific literature are shown below.

Students' concerns about reading scientific literature

- comprehension versus speed of reading
- coping with too much to read
- reading about topics with which you are not familiar
- understanding jargon
- making meaning out of complex data in tables, graphs, or statistical analyses
- sorting out which information is the most relevant
- understanding sophisticated scientific techniques or research methodologies
- dealing with contradictory information or points of view.

Many of these problems can be overcome by the use of appropriate strategies, with the further development of your own reading skills, and over time. Like most things, your confidence will grow with experience. You must remember that your inability to read and comprehend a particular piece of scientific writing may be due, at least in part, to your unfamiliarity with the style of writing in the sciences. Science writing is impersonal, uses passive verbs, and is often interspersed with data, graphs, and terminology that need to be interpreted as you read. Writers of science literature may not express themselves clearly or concisely. Perhaps they have not presented a simple and logical argument, which you the reader can follow easily. One of the assumptions many students make is that the writer, being an expert in their field, must always be correct. This is not always the case. You must learn to appraise objectively the work of others and not be swayed too easily by the data and the arguments presented by the writer.

The rest of this chapter will present a series of strategies and ideas which you may find helpful in formulating an approach which will help you get the most from the time and effort which you spend reading. It is not designed to be a foolproof recipe, but more of a starting point from which you can build your own methodologies.

TIPS FOR READING

In everyday reading different styles are used for different purposes. Flexibility of approach is much more important than absolute speed of reading. You may use any of the following techniques:

Scanning

You can scan a textbook or journal for a particular chapter or section, and you scan the table of contents or index looking for a particular topic.

Skimming

You can skim through a journal article for topics of interest, or through headings, tables, and diagrams looking for key data.

Reading

Normal reading can be any extended narrative retaining only those elements you need to follow the story.

Studying

When reading for a set purpose, you may read a textbook closely, going back over passages to understand their meaning. You may also highlight or underline important information, or make written notes and summaries.

These techniques are *all* relevant to reading scientific literature. Have a purpose for your reading, and, before you begin, write down a series of questions, for example:

*   Why am I reading this?
*   What specific information or data am I looking for?
*   How thoroughly do I need to understand all the information?

Then look through the piece of reading and determine what the big picture is:

*   *Scan* the table of contents.
*   *Skim* the headings, abstracts, and tables of results.
*   *Jot* down key words or phrases.
*   *Read* the summary or abstract.

Once you feel the content is relevant to your needs should you read the text more carefully.

Asking specific questions

In order to analyse a text effectively, you need a series of specific questions to which you seek answers. For example, say that you are reading a biology textbook chapter with the purpose of understanding all about alcohol dehydrogenase, the liver enzyme responsible for breaking down alcohol in your body. In order to focus your reading you may write down the following questions:

*   What is an enzyme?
*   How does the enzyme like alcohol dehydrogenase work?
*   Why is this enzyme important?
*   How does it relate to liver function tests?
*   What happens when this enzyme is not functioning?

As you read, make summaries using diagrams, with clear headings, *as you go* through the material. Look for the key ideas. With practice your notes should form the basis of the answers to the questions you formulated before you commenced reading.

Put the information into an order of priority in terms of answering the questions you set out to answer. Make a priority list for any further reading.

TIP

Do not fall into the trap of <u>underlining</u> or *highlighting* large amounts of the text. Read whole paragraphs first. Then go back and highlight the main ideas or most important pieces of information that relate to the point the author is making. The context of the whole text should be clear from the passages that you have highlighted.

Next you should summarise all the critical information in a format that you understand. That is, rewrite the author's work in language that makes sense to you, without distorting the facts or the essence of the argument. If necessary, make diagrams or 'road maps' of relevant information. How are these critical pieces of information linked? They may be linked by an equation, a formula, or a theory: attempt to put in those links.

Make your notes brief, say as much as you can in as few words as possible, and use diagrams. If necessary, have a good scientific dictionary at hand, which will enable you to look up terminology with which you are not familiar. The more you strive to understand what you are reading and why you are reading it, the more you are likely to remember.

Critical reading of journals

Journal articles are written by researchers in a particular field and are generally intended to be read by other researchers. Do not let this put you off. Good writers will make their meaning clear to you and have the ability to present complicated information or data in a simple and concise way. The following is intended to be a guide to reading a scientific journal article, with a specific purpose in mind, for example for an assignment or laboratory report.

Reading research literature requires two types of reading. The first involves a quick read, looking for clues as to the relevance of the article for your specific task. The second is a more detailed analysis of the information presented in the article, after having decided that it is relevant.

Skim reading
Title, author

The title of the article in itself should tell you if the contents are related to the information you seek. This may, however, not always be so. If in doubt read the abstract.

Abstract

- This is a short, concise synopsis of the rationale for the study, the main results, and their interpretation.
- This should be the first thing you actually read.
- If it sounds relevant, go to the summary or conclusion; if not, don't read further.

Summary and conclusions

- What are the main experimental findings and how has the author interpreted them?
- What are the main conclusions drawn from these findings?
- Is this information relevant to your specific task? If yes, then you need to read the article in depth.

Critical reading

Only after you have satisfied yourself that the details of the article are worth exploring should you attempt to read the whole text critically. In order to help you make meaning of the details it is again recommended that you attempt to answer a series of questions for each of the main sections as follows:

Introduction

- What is the overall aim or desired outcome of the study?
- What specific problem is being addressed?
- How is this study relevant to the specific problem?

Materials and methods

- What experiments, trials, or observations have been performed?
- What specific methods were used?
- Why were these methods employed?
- Under what conditions and limitations were the methods employed?

Results

- What were the main findings in the study?
- Are all results presented or only a representative sample?
- Look carefully at all the tables, graphs, and diagrams and form your own opinion as to what they mean.
- Are statistics presented? If so, are the statistical interpretations accurate and are they based on the experimental results?

Discussion

- Does the discussion summarise and interpret the results?
- Are these interpretations warranted based on the results presented?
- Are there any assumptions that have been made?
- What conclusions have the authors drawn from the results?
- Do the results of the study, their interpretation, and the authors' conclusions answer the aim of the study?
- Has the study contributed to an increase in understanding in the field?
- What further studies should be performed?
- How may the results of the study be used in other related fields?

You may not initially be able to answer all these questions. You may need to read the article several times. If you are having difficulty understanding particular aspects of the experimental techniques, the data, or other aspects of the paper, ask one of your lecturers or tutors for help.

Summary

Reading scientific literature with its specific terminology, use of language, and unfamiliar experimental techniques is a difficult task. The more of this type of literature you read, the more familiar you will become with the language, style, and terminology of the material. With time your reading speed, comprehension, and critical appraisal of the contents will improve. Be patient and persevere.

8

Critical Thinking

The history of science knows scores of instances where an investigator was in the possession
of all the important facts for a new theory but simply failed to ask the right questions.

Ernst Mayr, evolutionary biologist (1904–2005)

Key Concepts

- What is critical thinking?
- Science and the pursuit of knowledge
- Knowledge, scepticism, and critical thinking

- Why be a critical thinker?
- How to be a critical thinker
- What is justification?
- Critical thinking and argumentation

Introduction

The positive benefits of acquiring critical thinking skills are universally recognised by governments, business executives, scientists, academics, self-help gurus, and students alike. Yet it is not always clear what they mean by critical thinking. Sometimes the term critical thinking is used as a synonym for 'lateral thinking,' a term made famous by Edward de Bono. Lateral thinking is often described as 'thinking outside the square', which means thinking about a problem from a different angle or perspective in the hope that this will generate a solution or novel approach to a problem. Problem-solving of this kind is of great value to industry and the sciences but it is only one aspect of what it is to be a critical thinker. Not all the thinking we do is about solving problems, and that is particularly the case in academia. Another way that people view critical thinking is as a form of creative thinking. Again, while creative thinking may use some critical thinking skills, the terms are not synonymous. Creative thinking often leads to new insights or perceptions; it often leads to new products and ideas. Critical thinking, on the other hand, is not necessarily creative.

There is no general consensus on what critical thinking means but there is broad agreement about what it entails. According to the National Council for Excellence in Critical Thinking, critical thinking is:

> the intellectually disciplined process of actively and skilfully conceptualising, applying, analysing, synthesising, or evaluating information gathered from, or generated by, observation, experience, reflection, reasoning, or communication as a guide to belief or action.

Other definitions claim it is 'self-directed, self-disciplined, self-monitored and self-corrective thinking' (Foundation for Critical Thinking 2001) or 'the use of those cognitive skills or strategies that increase the probability of a desirable outcome' (Halpern 1997, p. 4).

While not identical, the definitions have some things in common. First and foremost, they all agree that critical thinking is about becoming a better thinker, becoming someone who not only thinks carefully and critically about what they see and hear around them but also someone who has the capacity to critically examine and reflect on one's own thinking. Second, they all agree that you engage in critical thinking to guide action, inform beliefs, gain knowledge, increase understanding, or help in decision-making. And third, they all agree on a core set of characteristics that include:

- analysis
- interpretation
- inference
- explanation
- evaluation
- self-regulation.

The following will explain what these characteristics mean and how to apply them to your studies. We will then look at a few tools to help develop your critical thinking skills.

Science and the pursuit of knowledge

> Very few really seek knowledge in this world. Mortal or immortal, few really ask. On the contrary, they try to wring from the unknown the answers they have already shaped in their own minds … To really ask is to open the door to the whirlwind …
>
> Ann Rice, The Vampire Lestat, p. 413.

The origins of critical thinking are deeply embedded in classical philosophy and the history of science. Classical Greek philosophers like Socrates, Plato, and Aristotle were the scientists of their day. They wanted to find out about the world they lived in; they wanted true knowledge that was immutable and absolute. At the same time, they realised that reality was not easy to find. Appearances could be deceptive and mistakes could be made. So they tried to devise a method that would guarantee truth and provide a strong foundation for knowledge. In particular they admired the mathematicians of the day, such as Euclid who had devised a series of formal geometric proofs that were seemingly irrefutable (we still learn Euclidean geometry at school). Their attempts to devise linguistic proofs akin to those found in mathematics were the beginning of what we call formal logic.

Critical thinking is also linked to the 'scientific method', an approach to science credited to Francis Bacon (1561–1626). The pursuit of science is about discovering new things, and determining the laws that govern them. It is also about providing explanations and making accurate predictions. It is about increasing our knowledge and understanding of the world, both for its own sake and to enable us to manipulate the world for our own ends.

Sir Isaac Newton is considered one of the greatest scientists of all time. He epitomised what is known as the scientific method. Unlike earlier scientists, he set up carefully crafted experiments to test his hypotheses and their predictions. He tried to prove his claims by using rigorous scientific methods of observation, hypothesis, and experimentation. The success of Newtonian mechanics, its ability to produce verifiable (provable) predictions, led to the widespread acceptance of the scientific method as a means of gaining reliable knowledge of the world. It was a way of providing reliable proof or evidence to support one's claims.

So there are three very important elements to critical thinking that emerge from its history, namely the use of:

* logic
* argumentation
* proof or evidence to support one's ideas.

Critical use of all three should enable you to discern what statements are most likely to be true or at least probable, what statements there are good reasons to doubt, and which statements we can't yet evaluate.

Knowledge, scepticism, and critical thinking

As with the first Academy, set up by Plato in 500 BC, the aim of universities is to make discoveries about the world in which we live and to increase our understanding and knowledge of that world. In the sciences, we focus primarily on developing our knowledge of the physical world and the laws that govern it.

> The subject matter of science exists independently of our knowledge of it, and the purpose of science is to describe and explain both observable and unobservable phenomena.
>
> Cambridge Dictionary of Philosophy (1999)

All good research attempts to contribute to our knowledge base by claiming something new or confirming previously made claims. Research attempts to establish what we call knowledge claims, such as 'plants metabolise carbon dioxide' or 'water boils at 100°C'.

These claims can be either true or false. In order to count as knowledge, the statement must satisfy the following criteria. It must be:

- believed (knowledge must be known by someone)
- true (it is a fact about the world as far as we know)
- justified (there should be convincing reasons for believing it to be true).

Example

It may be the case that there really is life on other planets, but it does not count as knowledge unless it can be proved, that there is enough evidence to support the claim. So the claim remains an opinion or a belief until we establish enough evidence for it to be convincing; only then does it become accepted as fact.

Despite the successes of science, there are some well-known problems with establishing ideas as knowledge. It is generally accepted that the majority of our knowledge claims can never be 100% certain. We cannot rely on perception to give us an accurate picture of the world because it is too easy to be mistaken. It was once thought that the Earth was flat and that the sun travelled around it, based on observation. Not only is our perception limited and prone to error but what we can actually perceive is limited to 'proximal' stimulus, that which can be directly observed or experienced. So even if our perceptual apparatus were foolproof, it would only give us knowledge about what is directly in front of us. How can we generalise to other similar things from such a small sample? How can we know about the things we can't see? This is a problem.

At the same time, most of the more important claims we make (or want to make) are about things we can't directly perceive even if they were in front of us, and these are called 'unobservables'. In the sciences, examples are atoms, gravity, and radiation. How do we know that these exist, or that the claims we make about them are true (figure 8.1)?

Figure 8.1   Realist vs sceptical view of science

Realist view	Sceptical view
The objects of science (like the atom or neutrons) must exist and the terms of theories refer (at least partially) otherwise the success of science is a miracle or just a cosmic coincidence. This is unlikely given the outstanding success of science and technology over the last few centuries. (Couvalis 1997)	The entities science posits play a useful role in theories but do not necessarily exist. Science has a history of being wrong or mistaken about the claims it has made about the world. Phlogiston turned out to be oxygen and hydrogen. Einstein showed that Newtonian mechanics was often wrong and that his own theory was 'simpler and more experimentally satisfactory'. (Couvalis 1997)

Today the most important criterion for knowledge claims is justification, the evidence, proof, or reasons one has for supporting a claim. The stronger or more convincing that evidence, or the more logical or persuasive the argument, the more likely it is that the claim is probably true. As scientists, we need to make sure that the claims we make are justified, that there is enough good, strong, reliable evidence to support those claims. We need to make sure that we have looked at alternative explanations of the evidence. What we aim for is the best explanation, given the available evidence (often called 'inference to the best explanation'). Applying critical thinking skills to the claims people make will help ascertain whether or not they are presenting the best or most reliable explanation given what we currently know.

Why be a critical thinker?

There are a number of very good objective reasons why critical thinking is a necessary component of academic work and a valuable skill to develop. Universities give you access to a rich source of information in an abundance of different media. This can be overwhelming. How do you know which information is reliable or trustworthy? How do you compare the quality of one source with another? At the same time, the more interesting books and articles use that information to make a claim that is not obvious. They will be putting forward a case or argument for a particular interpretation of the facts that they want you to find convincing. You need to have tools for assessing the relative strengths and weaknesses of the arguments presented.

You will also find writers who express different views and even conflicting or contradictory ideas about similar things. Who is right? How do you sort out which ones to accept and which ones to reject? You may find that the ideas expressed by some researchers are contradictory to your own personal thoughts, ethics, and ideals. Some of these ideas may conflict with your cultural knowledge and values. This can be difficult to accept. In these circumstances, you need to ensure that you deal with the material as objectively as possible. Even though you may not agree with a claim or find the conclusions go against your beliefs, you must judge the claims on their own merit. Developing critical thinking skills will help you do this.

Finally, you will need to construct and defend your claims in your academic work. The strength of the claims made in the texts you read is relative to the accuracy or quality of the evidence used to support them. The same applies to your own work. When writing, you need to show evidence that you have engaged in the critical process, assessed a variety of texts, and come to a well-reasoned conclusion. Reasoning critically will help you decide as objectively as possible about the quality of the information you encounter—by assessing and evaluating all the relevant issues and arguments.

Why critical thinking is important

- Books, lectures, and articles contain more than information. You need to sort fact from opinion or interpretation. Assess the argument.
- Writers express different or contradictory views — you need to compare and contrast.
- Can express bias or limited cultural perspective — you need to assess for objectivity.

- Ideas can conflict with personal beliefs, ethics, and knowledge – you need to be personally objective.
- Knowledge claims rely on the strength of the evidence used in support – you need to assess justification or proof.
- In constructing your own claims, you need to be clear, convincing, and logical.

How to be a critical thinker

Critical thinking in the sciences tends to focus on *problem solving*. However, we also need to critically assess how well scientists present their ideas and how rigorously they defend their conclusions. This will involve *text and research data analysis*, that is, evaluating how information is presented and interpreted. Text analysis primarily involves identifying and evaluating the justification for a claim or claims, that is, assessing the arguments. Within this context, critical thinking requires the skill or ability to:

- elicit the relevant claims and identify the justification
- analyse and evaluate the information presented
- make a well-reasoned assessment based on the information or conclusions drawn from the arguments.

What is justification?

Despite the difficulties of absolute certainty, there is an accumulated body of widely accepted scientific knowledge. There is a long history of scientific research and debate behind the development of that body of knowledge. It is legitimate to use this well-accepted knowledge as evidence to support your arguments. In fact, this knowledge is essential in understanding your field and in generating new ideas.

Observational evidence is another good source of knowledge, even though there are problems with certain kinds of observations in terms of reliability. We also need to bear in mind that observation alone can only tell us so much; the kind of knowledge we can draw from it *in isolation* is limited to singular statements about specific things. Although these sorts of facts are useful, and often infallible, they don't tell us much about the world. They mainly tell us about individual phenomena that we can perceive directly. However, we do rely on skilled and careful use of observations to provide us with useful information, particularly in the laboratory or out in the field.

As students and researchers, we use our own and others' carefully recorded observations, along with the accumulated body of knowledge to support the claims we make and to justify the inferences we draw. An important part of critical thinking involves a careful assessment of such claims and an evaluation of the evidence used to justify them. Figure 8.2 represents different kinds of justification that are used to support knowledge claims.

Figure 8.2   Justification of knowledge

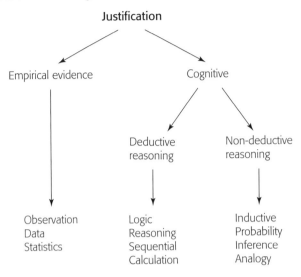

Assessing evidence

We use empirical evidence to support and justify the claims we make. It can be used to:

- identify or explain a phenomenon
- draw a conclusion or develop an hypothesis
- justify or prove an hypothesis.

In the sciences we tend to use empirical evidence, which may be observational, statistical, or experimental. All forms of evidence have their own criteria of acceptability. Once we have identified it, we can assess how well it fits its criteria. We can assess how reliable the evidence is, in relation to the justification used and the claim being made. The quality and quantity of the experimental evidence or physical data we gather is particularly important. It leads us to make claims about the phenomenon under investigation. These claims can have an impact on how we deal with that phenomenon so we need to be very careful about what it is we claim we have evidence of or for. We do this by asking ourselves questions:

- What is the evidence showing us? What can we really infer?
- Is the evidence reliable? How was it collected?
- Does it support our claims?
- Is the evidence enough? Is it relevant?
- Do the findings reflect or support those of others?
- Are there any anomalies?
- Have we overlooked something?
- Could there be another interpretation?
- What are the limitations?

Example of evaluating relevant evidence

In the mid nineteenth century, there was a particularly severe outbreak of cholera in London. It was widely believed at the time that cholera was an airborne disease and that the outbreak was caused by the extremely putrid condition of the air. A major engineering project was undertaken to improve the quality of the air by channelling London sewage out of the city in the belief that this would also remove the offending cholera bacteria. On completion of the project, the incidences of cholera disappeared, thus confirming the fact that cholera was indeed an airborne disease.

Ten years later, the cholera returned. Further investigation discovered that eels had been trapped in some of the Thames water pumps and died, contaminating the drinking water. Once these were sealed off from use, there were no new outbreaks of cholera. It turned out that cholera was, in fact, a water-borne disease.

It is easy to make mistakes, jump to the wrong conclusion, or jump to conclusions too quickly. Developing critical thinking skills by learning how to assess the accuracy, reliability, and quality of the evidence presented can help reduce error.

Critical thinking and argumentation

The first stage in analysing a text is being able to recognise an argument. Once we know what the argument is, we can assess and evaluate the strength of that argument by looking at its logical structure, the justification or evidence, and the appropriateness of the language in supporting the claim.

The easiest way to start thinking critically about what you read is to ask yourself questions:

- What is the author claiming?
- What is the argument?
- How is it justified?
- What evidence is presented?
- Does the evidence support the claim?
- Is the evidence convincing?
- What are the limitations of the evidence?
- Does the conclusion follow from the premises or evidence?
- Is the argument logical?
- Could there be another conclusion or interpretation?
- Are there other works which support these findings?
- Are there other works which contradict these findings?
- What assumptions has the author made? Are they valid?

- What is my view?
- Do you agree with the author? Why?

Once you have analysed the argument, you can now try to evaluate or assess its claims. You can do this in two ways—check the logic of the structure and the implications of the language used, and check the quality and quantity of the empirical evidence and its relevance to the conclusion.

Evidence is presented to support an argument. The stronger the evidence, the more convincing the argument. But remember, the evidence alone does not argue for anything; it does not speak for itself. We interpret the evidence and make a claim based on that interpretation or we draw an inference from the evidence. To do this, we rely on logic and what we know about the world. We make claims about what we think is the most likely explanation or interpretation of the evidence. And this is where we can go wrong.

Summary

Science has been described as the 'search for solutions', that is, recognising problems and then thinking rationally and critically about possible solutions to those problems. So perhaps the greatest skill that any scientist can have is the skill of critical thinking, which requires comprehension, critical analysis, evaluation, reasoning, and objectivity.

9

Academic Argument

We are entitled to make almost any reasonable assumption, but should resist forming conclusions until the evidence requires that we do so. Unfortunately, men and women rarely agree on what is reasonable or what counts as enough evidence.

Steve Allen, American comedian (1921–2000)

Key Concepts

- What is an academic argument?
- What makes a good argument?
- Deductive argument
- Inductive argument
- Using good argument in your writing

Introduction

In order to critique the texts you read, you need to understand what they are about, and what they are saying. While an article or book chapter may contain useful information or describe the way something works in great detail, there is usually a reason why that information is there. More often than not, the authors will be using the information to make a claim of some kind that they want you to believe. They will be putting forward an *argument*.

In your studies there will be times when you will be asked to put forward an argument. You will be asked to evaluate some experimental data and to come to a conclusion and then justify or argue for your claim. In order to be able to construct a good argument, you need to know what constitutes a good argument.

What is an academic argument?

In everyday language the term 'argument' is used to describe a dispute or disagreement. Within the academic world, however, the term argument does not always indicate a disagreement. An argument can be used to:

- support something we think has merit (a perspective, an experimental procedure, a piece of research)
- persuade someone that something would be beneficial to do (fund a project, take a particular course of action)
- convince someone that something is true, likely to be true, or probable (a theory, an hypothesis, an outcome)
- show someone the problems or difficulties with something (a theory, an approach, a course of action)
- reason with someone to get them to change their view or their practice.

   In its most basic form an argument is a claim that is justified. The claim (or conclusion) is supported by *at least* one statement, referred to as a *premise*.

What is an argument?

An argument IS ...

- A group of statements of which one is a proposition or claim that is supported by at least one of the other statements:
  Drinking water daily is good for your health **(claim)** *as it cleans out your liver* **(premise 1)** *and reduces the level of toxins in your blood* **(premise 2)**.

An argument IS NOT ...

- A statement of fact
  In 1996, 47% of all females in custody in Australia were Aboriginal.
- An assertion or claim
  Wearing a seatbelt reduces the risk of injury.
- A prescriptive statement
  The government should spend more money on healthcare.
- A conditional statement
  If you drink too much alcohol, you will damage your brain.
- A definition
  Analysis is the process of breaking up a concept, proposition, or fact into its simple or ultimate constituents.
- An explanation
  I was late home because my car wouldn't start.
- A series of statements about the same thing
  Most plants need plenty of water. Adding nutrients and aerating the soil can help them grow.

When do you need to present an argument?

You need to use argument when you want to establish a point or convince someone of a particular interpretation of the facts. In the sciences, you will mostly work with data from experimental observations from which you will be expected to draw reasonable conclusions.

- What do the results show?
- Are they what were expected?
- How should they be interpreted?
- What conclusions can be drawn from them?

Once you have come to a conclusion, you will be expected to justify it. You will be expected to give good reasons to support your claim. You will have to give reasons why you believe *x* or *y*, rather than *z*. The more convincing the reasons, the stronger the argument.

What makes a good argument?

A good argument should be convincing. You should find yourself believing the claim, or at least finding the conclusion reasonable. You should find it convincing based on the quality of the data and argument, not just the persuasiveness of the language. It should be convincing because there are very good strong reasons to accept the conclusion as probable. At a basic level this entails:

- *acceptability*—the premises are acceptable or reasonable (likely to be true)
- *relevance*—the evidence or reasons are relevant to the claim
- *grounds*—the reasons provide sufficient grounds to lead us to accept the claim.

These are called the *conditions* of an argument, and when they are satisfied it is likely to be a good argument. While the evidence is important support for a strong argument, the logical structure also plays a crucial role in its strength or weakness. If the language of the argument does not indicate the relevance of the evidence to the claim then it won't be convincing. And it won't really constitute an argument. A good argument requires both the right form (logical structure) and the right content (true premises).

Types of argument

There are two fundamental types of simple argument—*deductive* and *inductive*. Most arguments take one of these forms, though, in complex arguments, both kinds may be present. The difference between the two kinds of argument has to do with the logical relationship between the premises. Deductive arguments can be *valid* or *invalid*, while inductive arguments can only be assessed as *strong* or *weak*. If you use either of these forms you should ensure that your argument is either valid (if deductive) or strong (if inductive). How do you do this?

Deductive reasoning

We often admire people for their powers of deduction. Probably the person that most often comes to mind in this regard is Sherlock Holmes, the fictional detective. So what is deduction? In short, deduction can be defined in the following way.

> The concept of deduction is a generalisation of the concept of proof. A proof is a finite sequence of sentences each of which ... follows from preceding sentences in the sequence by a rule of inference. The relations between premises are syntactic not semantic, i.e. based on structure not meaning.

> (Cambridge Dictionary of Philosophy, p. 212)

Deductive reasoning is best represented by formal logic. The classic example is the three-step syllogism, put forward by Aristotle.

Example of deductive reasoning

All As are Bs

This is an A

Therefore, this is a B

Deductive reasoning involves constructing statements in such a way that the conclusion *necessarily* follows from the given premises (validity). Generally, most of the relevant information will be contained in a general statement that sets out the logical relationship between two or more sets of facts or events; an additional relevant fact or event will be presented and this will enable a logical deduction to a specific conclusion.

Example of a valid deductive argument

If x is an acid, it will turn litmus paper red.	$(p \rightarrow q)$
The litmus paper did not turn red.	$(-q)$
Therefore, x is not an acid.	$(\therefore -p)$

If the premises of a deductive argument are true and it takes a valid form, the conclusion is always true (what we call a 'sound argument'). It is not possible for the conclusion to be false, given the conditions set out in the opening premise, and the evidence presented in the second premise. The logical connectives (if ... then) stipulate the conditions under which the consequences occur.

If an argument is sound, we must accept the truth of the conclusion, as it is entailed by the logical structure of the argument. Soundness has to do with truth plus validity. Validity has to do with logic. If you follow the rules of logic, you can deduce certain inevitable conclusions. Beside the one above, standard forms of validity include:

Example

A or B	A then B
not B	B then C
therefore A	therefore A then C

The reason the argument forms are written using letters is because it doesn't matter what the specific content of a premise is. If an argument has the right form, then it is logically valid. If it is not valid, you don't have to accept the conclusion.

At the same time, we are only forced to accept the truth of the conclusion of a valid argument if the premises are true as well; that is, it is a sound argument. If one or other of the premises is false, we do not have to accept the conclusion. The argument is *valid* but *not sound*.

Example of a valid, unsound, deductive argument

If you don't have agriculture, you cannot develop a civilisation. (true)

Ancient South American Indians did not have agriculture. (false)

Therefore, they could not have developed a civilisation. (not necessarily true)

Example of an invalid deductive argument

If Newton's physics and requisite background assumptions are true,

then a comet will appear in December 1758. $(p \rightarrow q)$

A comet did appear in December 1758. (q)

Therefore, Newton's physics is true. $(\therefore p)$

Confirming the consequent (q), rather than the antecedent (p), is invalid as there could be other reasons why the consequent did or did not occur. For instance, it could be just coincidence that a comet appeared at that time. Or there may be another theory that could have predicted the comet just as readily. So the conclusion doesn't necessarily follow, it isn't inevitable. There are other possibilities. However, for *scientific theories*, confirmation of predictions (in this case, the consequent) is the only way to test their probable truth.

Using a deductive argument in an assignment can strengthen your claims as it enables you to draw true conclusions. This makes it a powerful tool, particularly in the sciences where we often have some reliable information already. For example, if we already know that x is either hydrogen or oxygen, and we find out that x is not oxygen, then we can deduce that x must be hydrogen. The structure of deductively valid arguments makes it impossible for the premises to be true and the conclusion false.

The conclusions you can legitimately draw from deductive arguments are often limited to specific cases or instances of things under very specific conditions. While useful in certain circumstances, they don't really tell us anything we don't already know; they don't increase our knowledge of the world. This is why the most common argument form is based on *induction*.

Inductive reasoning

Inductive arguments increase our knowledge, by extending what we already know and by seeing what can be inferred from existing knowledge. They are the most common and most useful form of argument and the type most used in the sciences. For example, the fictitious detectives in the television drama series *CSI: Crime Scene Investigation* would take a collection of very specific pieces of information, the evidence at a crime scene, and work out what the most likely course of events had been and what the most likely cause of the crime was. They would come up with the best, most probable explanation that fits the data. So what is inductive reasoning? Induction can be defined in the following way:

> In the narrow sense, induction is inference to a generalisation from its instances. In a broader sense, it can be any inference where the claim made by the conclusion goes beyond the claim jointly made by the premises (by analogy, predictive inference, inference to causes from signs and symptoms, probability, statistical inference, confirmation of scientific laws and theories).

(Cambridge Dictionary of Philosophy, p. 425)

Inductive reasoning is almost the opposite of deductive reasoning. We start from the specific instances and form an inference, either a generalisation or a prediction that seems reasonable or probable. In deductive reasoning we start off with our generalisation, or known condition, and deduce from it what we can, a specific fact or instance, based on the rules of inference.

The conclusion or claim of an inductive argument goes beyond the evidence or what is contained in the supporting premises. There is always the chance that the conclusion could be untrue. We can say that 'All metals expand when heated' is an inductive generalisation based on a limited number of observations. We can't test all metals under all conditions. There may be exceptions. Because of this, an inductive argument is always invalid and the truth of the conclusion can never be guaranteed based on the argument. Nevertheless, you can construct a strong inductive argument, where the conclusion or claim is considered highly probable or extremely likely. The strength of an inductive argument is dependent on the amount of supporting evidence relative to the claim. Simply put, strong evidence leads to a highly probable conclusion, weak evidence leads to an improbable conclusion.

Example of an inductive argument

Consider the following inductive argument where a generalisation can be made based on the repeated occurrence of a particular event.

> There are many instances where drinking too much alcohol is linked to violence.
>
> Therefore alcohol makes you aggressive.

The conclusion or claim is not deduced but inferred from the evidence presented in the premise. The conclusion is not a necessary consequence of the premise. We could have concluded that 'alcohol use should be restricted' or that 'more security needs to be present at venues that serve alcohol.' We may also need to ask 'Does alcohol always make everyone aggressive?' So we might want to qualify our argument and say that alcohol *can* make you aggressive. This weakens the strength of the claim but does not weaken the strength of the argument. In fact, it could make the conclusion more believable or persuasive.

This is one of the major differences between inductive and deductive reasoning. When we use inductive reasoning, we try to see where the evidence leads us, what kinds of things could we reasonably draw from it, or how it could be interpreted. We judge the strength or reasonableness of the claim based on the quantity and relevance of the evidence used as support.

When writing your assignments you need to think about what you can justifiably claim and what you cannot. What does the evidence support and what doesn't it support? Be careful to only claim what can be reasonably inferred from the evidence. Qualify your answer if necessary or back up your claim with stronger, more reliable evidence.

What makes a weak or strong inductive argument?

Look at the inductive arguments below. The claim or conclusion is *not* an inevitable consequence of the premises and could possibility be false. Nevertheless, some of the arguments are stronger than others because the reasons (evidence) are more reliable or more relevant or both.

Examples of inductive arguments

a   The weather is unstable at this time of the year and we have had a few days of sunshine.
   It is likely to rain this weekend.

b   I have yet to meet one person who knows anything about food irradiation.
   From this I conclude that no one is aware of the benefits of nuclear technology.

c   The sun has risen every day for as long as history has been recorded.
   The sun will continue to rise every day.

d   Nicotine is an effective pesticide because when it is sprayed on the leaves of a plant the plant resists attack by insects.

e   Four per cent of all cereals contain peanuts.
   Tom has an allergy to nuts.
   Tom will have an allergic reaction if he eats cereal.

Let us look at the reasoning used in these arguments and see if they are strong or weak.

Argument (a) is making a prediction based on past events. The assumption is that because things occurred in the past a certain way, then they will occur the same way in the future. Secondly, there is an assumption that the odds are in favour of rain, because there have already been a few sunny days. We know that weather patterns are not predicted in this way. This argument is not very strong.

Argument (c) uses a similar argument based on past events. There is no guarantee that the sun will continue to rise. The prediction is based on the regular and continuing movement of the planets in our solar system around the sun. So even though the conclusion may be highly probable, it doesn't mean the reason offered is the most convincing.

Argument (b) is an example of a sweeping generalisation, where the conclusion claims far more than the evidence warrants. How many people has the writer encountered? There will be people who do know the benefits of nuclear energy. Is food irradiation a benefit? What are the benefits of nuclear technology?

Argument (d) is a generalisation but one where there seems to be a greater degree of likelihood based on the evidence. It does not appear to be hasty or overgeneralising, given that it is only claiming something within the confines of the evidence (nicotine as pesticide).

Argument (e) uses probability to argue that Tom will have a reaction. Here we have to judge how likely that claim is, based on the odds. The likelihood of a reaction is probably too low to warrant the strength of the conclusion.

An argument using poor inductive reasoning when only limited evidence is used to justify a claim that is not warranted from the evidence is often referred to as a *fallacy*. For example: 'All Fords are rubbish because the one I bought broke down.'

This is a hasty generalisation. It involves drawing a general inference about a group of things from only one example. An overgeneralisation involves making claims about a whole class of objects based on certain characteristics of individuals in that class which may, in fact, not be shared by all objects in that class, for example racial or sexual stereotypes.

Improving an inductive argument requires the addition of more or better supporting premises, or a conclusion which is based only on the evidence at hand. Consider the following example:

Example

The new oral contraceptive drug Luvinol was trialled by 100 women in Australia for 12 months.

None of the women complained of side-effects.

Therefore, it is safe for women to use this drug.

The inductive argument presented here is weak because, though the conclusion could be correct, the evidence to support it is weak, and the claim too sweeping. A stronger argument is presented in the following:

Example

The new oral contraceptive Luvinol, was used for a period of twelve months by 5000 female volunteers, aged 18–40 years with no history of major health problems. No ill effects were reported during the trial period. Annual health checks for a further period of five years found no observable health problems among the trial group. This indicates that the drug Luvinol could be a safe contraceptive for healthy women.

It is important to note that the conclusion or claim of any of the above arguments may be true or turn out to be true. What counts is whether or not the evidence presented as support or justification warrants reaching that conclusion. That is our measure of the strength or weakness of the argument. How we judge the evidence or justification will depend to some extent on the context, on what is relevant or sufficient evidence for that particular thing in those circumstances. This is why we are more likely to accept the claim that the sun will rise tomorrow than the claim that it will rain on the weekend.

Using good argument in your writing

The main point in writing an argument is having something to say. There has to be some point you are making, some claim you want to justify or some position you want to put forward. Once you have decided what that is, you have to think how you can support that point of view using the evidence you have.

Be self-reflective. Ask yourself, what made me come to this conclusion? What makes me think this idea is right? What did I find out that was convincing? What was I testing and what were my results? What did I read and why was it persuasive? What did I find problematic? These will be the reasons that justify your claim or conclusion. If you found the evidence convincing, then so should the person reading your assignment.

Once you have sorted out what it is you want to say and why you want to say it, you need to work out how to say it as effectively as possible. This is where the logical structure of your argument becomes important. Presenting facts or evidence alone is not an argument. You need to state clearly what that evidence is showing and why it is relevant to your case. You need to pull out the implications. What does it indicate? Why is this important?

Make sure in written assignments that you present your reasons in a logical sequence. Deal with one idea at a time to make sure your point is made as clearly as possible. Additional evidence should add strength to your claim by a process of accumulation. However, you need to be sure it is reliable and indicate how reliable it is. Always remember to connect back to your original claim so that the reader doesn't forget what it is you are arguing for. Don't forget to address any arguments that work against you. If there is evidence that undermines your claim or weakens it you must address it.

Qualities of a good argument

- Your claim or point of view is clearly stated.
  - What will you be claiming or arguing?
- Your evidence supports your claims.
  - Include supporting evidence.
  - Acknowledge counter arguments or counter evidence.
  - Use appropriate language.
  - Evidence alone does not make an argument, it merely supports it.
- You draw out the implications.
  - Why are you saying this here?
  - What is the point you are trying to make?
  - What does this evidence show?
- Your writing has a clear logical structure.
  - The points are relevant.
  - They lead towards the conclusion.
  - You show clearly where you are heading and why.

Summary

Whether you are agreeing or disagreeing with an author, a point of view, or a theory, whether you are presenting an alternative approach or supporting someone else's, or whether you are giving a particular interpretation of experimental observations, you are expected to give reasons to justify your claims. You are, in fact, presenting an argument. The same criteria you use to critically assess and evaluate the work of others will be applied to your own writing. You will be judged by how well you present your evidence, how logically you present your ideas, how thoroughly you have done your research, and how you acknowledge and deal with alternative points of view.

PART 4

Writing
and Presenting
in the Sciences

10

Academic Integrity and Plagiarism

Have the courage to face the truth. Do the right thing because it is right. These are the magic keys to living your life with integrity.

W. Clement Stone, American author and philanthropist (1902–2002)

Key Concepts

- Academic integrity
- What is plagiarism?
- Examples of plagiarism
- How to avoid plagiarism
- Paraphrasing correctly

- The anti-plagiarism plan
- Group work
- Examples of unacceptable group work

Introduction

Research in Australian universities shows that students feel more competent and confident if they know what to expect and what is considered acceptable in their tertiary institutions. Each of you comes to university from a variety of learning backgrounds and cultures and it is up to each of you to accommodate to the learning culture of your university. One aspect of this culture that is common to all universities is the concept of academic integrity.

What may complicate matters is that different universities may have different rules they wish their students to follow. So with a variety of expectations present on coming to university, you may find some confusion about what is expected. You may be told to follow a certain procedure in one topic and other procedures in another topic. This chapter aims to help you to understand the broad concepts of academic integrity as it applies to *all* universities and will describe the meaning of plagiarism, paraphrasing, and working ethically in groups.

Why do we need academic integrity?

There are many reasons why we need to understand, use, and display academic integrity when we study at university. In an academic context we show consideration for our culture and demonstrate our professionalism by being honest and acting with fairness, respect, and responsibility.

Acting with integrity is part of the chain of academic learning. We must give credit where credit is due. There are various behaviours that compromise academic integrity. These include copying the work of others, collusion, cheating in examinations, fabricating or falsifying data or results, or not giving credit to the work of others in your own work. These are all regarded as serious breaches of academic integrity. One way to be certain of what is and what is not acceptable at your university is to consult its version of the *Student Related Policies and Procedures*, which is likely to be available online. All Australian universities should have a stated policy on academic integrity and plagiarism issues.

What is plagiarism?

A form of academic dishonesty, referred to as *plagiarism*, may be defined in the following way.

> *Plagiarism consists of using other person's words or ideas as if they were one's own. It may occur as a result of ignorance or inexperience about the correct way to acknowledge and reference authors. It may also occur as a deliberate misuse of the work of others with the intent to deceive.*

There are serious consequences for students who breach any of the policies of academic dishonesty, so make sure you consult your university's website and find the relevant policies.

Some common examples of plagiarism

- submitting someone else's essay (or any assignment) using your name as though it was your own work
- downloading text (essay, assignment, or any text page) from a website and using all or parts of this text in your work as though it were yours
- copying a section of an article from a journal or a book and submitting it as your own work
- copying sentences, paragraphs, diagrams, or pictures from someone else (essay, article, book, lectures) and using them, without proper acknowledgement to the source
- copying whole sentences or paragraphs from others and then changing a few words or rearranging the word order and then calling the work your own
- using a sentence or paragraph 'word for word' from another source, without the use of quotation marks and a reference to indicate that the words are not yours
- using the ideas and work of others and putting them into your own words but not saying where those ideas came from.

Other forms of unethical behaviour

The following are examples of student behaviour that would be considered as not complying with academic integrity:

1   Collusion, which is defined as presenting work as if it has been done independently when it has been the result of collaboration

    David, Rebecca, and Ion scored full marks on their online biology quiz, which they did together but which they pretended they did independently.

2   Copying the work of other students

    Bree felt sorry for Tam who had missed a class through illness, and allowed him to use her laboratory results for his assignment.

3   Purchasing or obtaining essays, tutorial, test or exam answers

    Sam paid for answers for the weekly quiz held in his tutorial on Tuesdays from Pierre who took the quiz on Monday.

4   Taking unauthorised material into an exam, sitting an exam for another student, or having another student take an exam in your place

    Ralph had gone terribly at his last exam and panicked that he wouldn't do well in the next so he asked his brother to sit the exam for him.

5   Making up references, or data, or giving secondary sources as if they were primary ones

    Simone worked extra shifts at her part-time job and felt she didn't have time to look up the original data for a science report so she used material from a secondary source that had written about the study but didn't include those details in her work.

6   Hiding library books or articles, or cutting out pages or deleting text

    Josh wanted to make sure that no one else in his course had access to case notes which had helped him answer an assignment so he hid the book in another part of the library.

7   Lying about medical or other circumstances to get extensions or special consideration

    Yoko asked for an extension on compassionate grounds that her grandfather had died when what she was doing was taking a trip on the Great Barrier Reef.

8   Having someone or an editor substantively edit or proofread a piece of writing.

    Winston worried his thesis argument might not be sound so he hired a professional editor and paid for his thesis to be substantively edited.

Referencing

At university we must acknowledge other authors' ideas and words. Scholarly writing usually demonstrates a familiarity with a range of writing and research. Handling our research writing properly means it adheres to academic conventions. You'll need to be accurate with citing sources in essays and assignments and also at referencing them fully and correctly. When someone is reading your academic writing they'll also need to recognise what is a direct quote and what is a paraphrase. Many science topics do not like the idea of students using direct quotes from written texts and discourage students from doing so.

Different systems of referencing may be used across disciplines on the same campus and there are certainly many different systems used worldwide. Generally, there will be a convention that suits an individual area of study. You'll need to consistently follow the conventions of *one* style of referencing with each piece of scholarly work. Check with your topic to see which style you'll need. Two of the most widely used referencing systems in the sciences are extensively covered in Chapter 11.

Why reference?

We reference material and ideas sourced from the work of others for a number of reasons:

1   *It is required by custom and, in some cases, even by law, to give credit where it is due.*

    Like the credits at the end of a film, the references and citations indicate who is responsible for the different elements that have been brought together to make the whole.

2   *We reference and cite to distinguish ourselves as authors of our original work.*

    The reader needs to know which ideas are your original contributions. The best way to do this is to indicate what is not yours (it should be considerably less than what is yours).

3   *We reference and cite so that readers who are interested in the subject can do further reading, or can verify interpretation of the evidence.*

    Interested or astute readers will be keen to know more. You may think in your first year of undergraduate work that nobody but your tutor will read your work, but you never know—anyway, you should get into good habits right from the start.

Paraphrasing correctly

To paraphrase is to take the work of others and put it into your own words. This is an important skill. The following shows examples of what is regarded as plagiarism and what is not. There is an explanation below each one.

Example

Example 1

In 2004, Professor Mike Lawson wrote about student teachers' vocabulary and what it said about their understanding of the teaching process:

> It is striking that the descriptions of learning processes generated by Cathy, and by many of her fellow students, are common language, rather than technical descriptions. When she is learning Cathy reads, understands, applies, draws upon, picks up, pulls together, and absorbs. These descriptions would not qualify as examples of what Elen and Lowyck (1999) referred to as 'systematic vocabulary about learning'. The descriptions are general

rather than specific in nature and do not depend upon the technical terminology associated with learning processes. Upon probing in the interview, the meanings of common language expressions could not be elaborated upon. Nor were the different descriptions explicitly interrelated to any significant degree and there was no articulation of a broad model that showed how different processes could be placed within an organising framework.

> Lawson, M. (2004) 'What students know about making good use of learning and teaching situations'. In K. Deller-Evans & P. Zeegers (eds), *Language and Academic Skills in Higher Education,* vol. 6, pp.11–24, Flinders University, Adelaide.

Student A wrote:

Greater emphasis is now placed on the processes of learning for those students who will become teachers, but this is not reflected in their own learning and teaching situations. Research shows that education students don't use a 'systematic vocabulary about learning' when discussing their own process of learning (Lawson 2004, p. 12).

This *is plagiarism* even though it was probably unintentional. The student has cited the source (Lawson, 2004) appropriately in the first instance, but in the last sentence has neglected to note that the quote used was from another source. This is an occasion where you need to use 'cited in' and cite the other reference. This may be an oversight by the student, or it may be that the student simply did not understand that the quote used by Lawson came from another source.

Student B wrote:

There has been greater emphasis on the processes of learning for those themselves going out to teach. Yet the vocabulary of students who are to become teachers does not reflect their own learning and teaching situations (Lawson 2004, p. 12). Research shows that education students don't use a 'systematic vocabulary about learning' (Elen & Lowyck 1999, cited in Lawson 2004, p. 12) when discussing their own process of learning.

This is *not plagiarism*. The student has cited the sources appropriately.

Example 2

The US Environmental Protection Agency wrote in its *Guidelines for Water Reuse*:

In an effort to help meet growing demands being placed on available water supplies, many communities throughout the US and the world are turning to water reclamation and reuse. Water reclamation and reuse offer an effective means of conserving our limited high-quality freshwater supplies while helping to meet the ever growing demands for water.

The student wrote:

Many council areas throughout Australia are using water reclamation and reuse. Water reclamation and reuse offer an effective means of conserving our limited high-quality freshwater supplies while helping to meet the ever growing demands for water.

This is *plagiarism* as the student has not cited any source, either for the direct quote of the second sentence, or the inadequate paraphrase of the first sentence. Merely substituting words in someone else's text does not create an effective paraphrase. Using research from overseas and passing it off as local is also inappropriate.

Example 3

In the journal, *Nanotechnology*, there was a news item about the use of nanoparticles to detect the presence of bacteria in meat.

> Researchers at the University of Florida, USA, have devised a nanoparticle-based bioassay that can detect a single bacterium within 20 minutes. They used the technique to search out *E. coli* bacteria in samples of ground beef. 'Highly sensitive and reproducible detections of bacteria and other biological agents will save lives, provide better health care, effectively fight terrorists and aid food safety,' said Weihong Tan of the University of Florida.
>
> Institute of Physics Publishing Nanotechnology 2004, 'Particles could beef up food safety', *Nanotechnology*, vol. 15, no. 12.

The student wrote:

> Researchers say nanotechnology represents the solution for many of the world's current problems, including unsafe food practices. Discoveries of the causes of bacteria in food will benefit the living, improve the provision of medical and related services, and wage war on acts of radicals.

This is *plagiarism*. The student neglected to cite the source of the information used. This student has merely substituted words from the original source, which is not an acceptable form of paraphrasing.

Example 4

Clemente et al. (2004) described how a lizard's movement style relates to its eating style.

> Australian dragon lizards (*Agamidae*) are typically fast-moving, sit-and-wait predators with a catholic diet consisting mostly of small invertebrates (Pianka 1986). In contrast, the thorny devil (*Moloch horridus*) is a highly specialised, slow-moving predator of small ants, primarily *Iridomyrmex* and *Crematogaster* (Withers & Dickman 1995; Witten 1993; Cogger 1992; Greer 1989). It is exceptionally slow moving, and often rocks backwards and forwards while moving slowly or when stationary (Witten 1993; Cogger 1992; Greer 1989).
>
> Clemente, C.J., Thompson, G.G., Withers, P.C. & Lloyd, D. 2004, 'Kinematics, maximal metabolic rate, sprint and endurance for a slow-moving lizard, the thorny devil (*Moloch horridus*), *Australian Journal of Zoology*, vol. 52, no. 5, pp. 487–88.

The student wrote:

> Australian dragon lizards are typically fast-moving, sit-and-wait predators with a catholic diet consisting mostly of small invertebrates. The thorny devil, *Moloch horridus*, is a predator of small ants but it is slow moving, and often rocks backwards and forwards while moving slowly or when stationary (Clemente et al. 2004).

This is *plagiarism*. The paraphrasing here is very inadequate as the student has almost used the words of the source verbatim but failed to use quotation marks around the quoted passages. The student has also failed to cite the secondary sources cited by the author for the observations about the thorny devil.

The text would look better like this:

> The Australian dragon lizard (*Agamidae*) is a fast moving predator which will wait for its prey. It has a wide ranging diet consisting mostly of small invertebrates (Pianka 1986, cited in Clemente et al. 2004). In contrast, *Moloch horridus*, the thorny devil, has been

described as slow moving (Witten 1993, cited in Clemente et al. 2004; Cogger 1992; Greer 1989) and is a predator of small ants (Withers & Dickman 1995, cited in Clemente et al. 2004; Witten 1993; Cogger 1992; Greer 1989)

A worked example

An approach to take when using information from any source is to follow a sequence of steps to ensure you are following the correct procedure. The following is suggested as an example. Consider that you are using the data from the following extract of a journal article as part of an assignment.

Asian Journal of Educational Psychology 2006, **29,** 80–90.

Social skills and learning approaches

Julius Marks

Hong Kong University

Sample

The subjects were 190 students in their matriculation year of secondary school in Kathmandu (Nepal). The subjects were 15 years to 17 years of age, with 90 males and 100 females.

Instruments

The Better Study Instrument (BSI) (Keaton, 2000) is an instrument that recognises six motive or strategy areas of learning (Surface Motivation, Surface Strategy, Deep Motivation, Deep Strategy, Achieving Motivation, and Achieving Strategy). It consists of 60 items, answered on a 5-point scale from: 1 = 'never true' to 5 = 'always true.' The measure of social skills utilised was the 20-item form of the Social Ability Scale (SAS) (Laurel & Hardy 1990). Both instruments were translated into Nepalese using the translation-back translation method. The BSI has been widely used in South-East Asia and provides considerable evidence of the reliability and factorial validity of the Nepalese version. Items of the scales were checked for relevance to Nepalese subjects and the scale itself was found to have an internal consistency reliability estimate of 0.81.

Results

The correlations between the SAS and BSI scales are shown in table 2 separately for males and females. The results suggest that the relationship between social skills and approaches to learning is relatively strong on the Surface Strategy ($r = -0.35$) and Achieving Strategy ($r = 0.40$) scales. There is also an indication that social skills are particularly related to low Surface Strategy scores for the boys but for the girls they are more related to higher scores on the Achieving Strategy scales.

Your research notes

- Study with Year 12 students in Nepal (N = 190, male = 90, female = 100)
- Approaches to learning measured using the BSI instrument
- Social skills measured with the Social Ability Scale (SAS)
- SAS shows a strong correlation with BSI scales, in particular:
  - SAS shows strong relationship to low SS score for boys
  - SAS shows strong relationship with high AS for girls.

Paraphrase 1: a simple summary statement

A study by Marks (2006) of final-year secondary students in Nepal found a positive correlation between measures of socials skills and approaches to learning scales. Marks further found some evidence of a difference between boys and girls.

Add the reference

Marks, J. (2006). The influence of social skills on learning approaches, *Asian Journal of Educational Psychology,* **29,** 80–90.

Paraphrase 2: a more detailed synopsis

Marks (2006) conducted a study of the relationship between socialisation, using the Social Ability Scale (SAS) (Laurel & Hardy 1990) and study approaches, using the Better Study Instrument (BSI) (Keaton 2000). Marks found that for final-year secondary students in Nepal there was a moderate correlation between socials skills and Surface Strategy ($r = -0.35$) and Achieving Strategy ($r = 0.40$) scales. Marks further found evidence of a difference between boys and girls on the Surface Strategy and Achieving Strategy scales.

Reference list

Keaton, A. (1997). The Better Study Instrument: a method of evaluating student learning, *Psychometric Evaluation,* **45,** 81–90.

Laurel, P. & Hardy, P. (2000). The development and testing of the Social Ability Scale instrument, *Educational Psychology,* **21,** 30–41.

Marks, J. (2006). The influence of social skills on learning approaches, *Asian Journal of Educational Psychology,* **29,** 80–90.

The anti-plagiarism plan for your own work

As a means of avoiding any possible hint of plagiarism in your own writing, the following approach is suggested:

1   Always give credit to another author's:
    (a)   exact words
    (b)   ideas, theories, hypotheses
    (c)   data, figures, tables, or research findings.
2   Give credit of source material in two places:
    (a)   in the assignment as an in-text citation
    (b)   at the end of the assignment in the reference list.
3   In order not to get caught up using too much by a particular author or article:
    (a)   have your own purpose and point of view
    (b)   use your own plan or outline
    (c)   remember your audience.

4   When authors' views conflict with each other, evaluate each author and the article and use that which you think is most reliable. To do this, think about:

(a)   how recently the material was published (generally the more up-to-date the better)

(b)   where it was published (for example compare *Nature* to *The Australian*)

(c)   whether the author is well regarded (you may not know this)

(d)   whether the work has credibility (for example has it been cited by other authors?).

5   Only use direct quotes when the precise words of the author are needed to justify your interpretation and quote in the correct manner.

6   Make sure you follow all the rules of the referencing system you are using.

Warning

A final few words of warning about plagiarism. Some universities may require you to submit a signed declaration stating that all the work you have submitted for assessment is your own, unless due reference is made to the work of others. Many universities today are now also using electronic anti-plagiarism programs such as *Turnitin, Edutie,* or *Plagiserve* to detect plagiarism in students' work when submitted electronically. So be warned, if you plagiarise, you will be caught.

Group work

Group work is a number of students committed to working together towards a common outcome or goal, using agreed processes, for which each member is accountable. Being able to work as a part of a team is highly valued in today's society. While you are at university you will probably have at least one group work project included as a part of your course. You still need to stick to academic conventions for group work, because you will be formally assessed on your work.

In your topic you should be given guidelines for assessment of individual effort within group work. There is a variety of ways in which you may be assessed, based on individual input, group-wide output, by peer or self-assessment, or combinations of these. If you are unclear about the expectations of a group work project, seek to clarify them with your tutor.

Legitimate collaboration involves constructive educational practice. Successful group work is supported by good preparation by members and clear ideas of individual tasks. It is important to have a common goal and excellent communication between all members of the group. What is not acceptable is a variety of behaviours that constitute cheating in one form or another.

Examples of unacceptable group work behaviours

1   *unauthorised* collaboration, called collusion

Jackie and Colin worked together on both their report chapters but then presented their reports as if the work had been done independently.

2 *copying* from other members

Mai mucks around in a practical class then realises later she hasn't taken down the results, so she asks Tim for his and copies his data to fill in the gaps.

3 *contributing less* of a share than other members of the group

Raj claims to have participated in the total group work experience but does not turn up for meetings or do the required percentage of independent work towards the total group output.

4 *contributing more* of a share than other members

Charles is a high achiever and is worried his group won't contribute enough or on time so rather than communicating with the others he does all the work himself and tells the others it's their share.

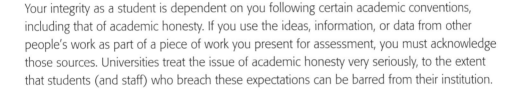

Summary

Your integrity as a student is dependent on you following certain academic conventions, including that of academic honesty. If you use the ideas, information, or data from other people's work as part of a piece of work you present for assessment, you must acknowledge those sources. Universities treat the issue of academic honesty very seriously, to the extent that students (and staff) who breach these expectations can be barred from their institution.

11

Referencing Styles

If I have seen further, it is by standing on the shoulders of giants.

Isaac Newton, English scientist (1643–1727)

Key Concepts

- Referencing terminology
- Referencing styles
- Author–date (Harvard) style
- Vancouver style
- Keeping track of your references

Introduction

Imagine for one moment that you are a research scientist working on a project to find a cure for cancer. After more than twenty years of work, you make a breakthrough. You discover a new compound isolated from the poison glands of an Australian marine jellyfish that selectively kills cancer cells in the lungs without damaging healthy cells. You call the new drug AMJ99 and with some further modifications to the original structure you synthesise other variations which selectively kill breast, pancreatic, prostate, and brain tumours. You become rich and famous, and before you know it you are in Oslo receiving the Nobel Prize for Medicine. At your acceptance speech, you thank by name the other scientists whose work and ideas helped shape your thinking and led you to make your breakthrough.

Well, so much for daydreaming, let's get back to reality. One of the most important skills you will develop at university is the skill of professional writing, which will allow you to convey succinctly and concisely your ideas to other professionals. Writing in the academic and professional world requires, in fact demands, that you let the reader know when you are using information and ideas from other people's work. In the scenario created above, you thanked those whose work guided you to your discovery. In written work, by citing your sources when you are writing, you are doing the same thing and, in so doing, you allow the reader to distinguish between other people's ideas and your own. You also give the reader of your work the opportunity to consult the references you cited or follow up interesting ideas. Most importantly, the reader knows how your thinking has been shaped by what you have read. This scholarly approach is what Newton was referring to in his statement at the head of the chapter.

Chapter 10 introduced you to the concept of academic integrity and some of the things that are required of you in the academic world. This chapter will be devoted to the methods you can use to inform the reader of your sources of information with the minimum interruption to the flow of your writing. This process is called *referencing* and is an integral component of academic integrity.

Key terms for referencing

Bibliography	a list of works consulted in preparing your text
Citation	the process of documenting (or stating) your sources
Paraphrase	material taken from a source, but put into your own words
Quotation	the exact words from a source, in quotation marks or set apart from the text
Reference list	the list of source materials from which you have taken information directly
Sources	other people's material taken from books, articles in periodicals, websites, journals, newspapers, lectures, television programs, or artworks

Referencing styles

There are many different referencing styles commonly used at university, so the style you are asked to use in one of your topics could well differ from that asked for in another topic. The most commonly used referencing styles in the sciences are:

- Author–date system (for example the Harvard system): the most generic system
- Vancouver system: used extensively in the medical and physical sciences
- Council of Biological Editors (CBE style): used in the biological sciences
- APA system (American Psychological Association): used mainly in psychology

Each of these systems has two common elements:
- the citation in the text
- the reference list.

Non-referenced material

Before we look at how to use the first two referencing systems mentioned there is one further point to make. There is much in the world that is well known, or that is accepted knowledge. In these situations it is not necessary to cite the original source of the information, which may in any case be obscure or difficult to determine. For example,

- Smoking is bad for your health.
- Speeding is dangerous.
- Overeating leads to obesity.
- Giving blood saves lives.

Author–date style

The following information on the use of the author–date or Harvard referencing style is based on that in the *Style Manual,* 6th edn, John Wylie & Sons, Australia, 2002. You may find that different universities, or even different topics at the same university, use variations of this style that differ in some ways (such as punctuation) from the examples cited here. It is thus *vital* that you check your topic guide or handbook for the exact style you are asked to use.

Citations in the text

When you write an assignment you must include, *in your text*, references to all the material you have used directly as source material for the content of your work. To identify references within your text, the author's last or family name and the year of publication of the material must be cited. Page numbers can be used when you need to direct the reader to specific information. Page numbers *must* be used when you quote or paraphrase a particular passage, or use specific data, graphs, or figures. Use *p.* for a single page, *pp.* for more than one page.

> *Smith (2006, p. 45) has argued that 'the relative seriousness of the two kinds of errors differs from situation to situation'.*

> *It has been argued that 'the relative seriousness of the two kinds of errors differs from situation to situation' (Smith 2006, p. 45).*

If you *paraphrase* material you can include a page number to make it clear where in the original text the information is located:

> *A recent study (Jones & Chan 2006, pp. 30–2) presented research data that …*

When the authors' names are incorporated within your text, you must write the word 'and', not the ampersand '&':

> *Jones and Chan (2006, pp. 30–1) have shown that …*

Reference to material written by *more* than *three authors* should include only the surname of the first author and the abbreviation *et al.* (meaning 'and others'). In the following example, the citation refers to work done by Lim, Smith, Brown, Smart, and Nguyen:

> *A recent study (Lim et al. 2006) has shown …*

Reference to *different authors with the same surname* should be distinguished by using the authors' initials or full names:

> *A recent study by Jones, C.L. (2006) has shown … but Jones, A. (2005) has suggested …*

When you have read an account of original work by one author (primary reference) in another book or article (secondary reference), both sources must be acknowledged in your citation. However, only the secondary reference (the book or article you actually used) should appear in your reference list. In these examples, Marini (2005) is the original work which is then cited by Jones (2006):

> *Marini (Jones 2006) states …*

> *Marini's study in 2005 (cited in Jones 2006) states …*

> *Jones (2006), in reporting Marini's study of 2005, states …*

If you need to cite several references at the same point, separate the authors' names by semicolons, with publications normally in *chronological* order. Citations published in the same year are listed in alphabetical order using the first listed author's name.

> *Recent studies (Brown 2004; Wong & Smith 2003; Kuwlesky 2002) have shown …*

References to two or more publications in the same year by a given author should be distinguished by adding a, b, … to the date.

> *Recent studies (Jones 2005, 2006a, 2006b) have shown …*

References to personal communications should include initials, last name, and date, followed by *pers. comm.* They must appear in the text, but not in the reference list.

> *These irregularities represent distal turbidities (J.A. Lethbridge 2006, pers. comm.).*

Reference list

The reference list is placed at the end of the written text. The list is arranged in *alphabetical order* of authors' surnames and chronologically for each author, with the items marked below with an asterisk (*) being essential.

When using the author–date system for *books*, the following information is required in order:

* author's surname and initials *
* year of publication *
* title of publication (in italics) *
* title of series (if applicable)

- volume number or number of volumes (if applicable)
- edition (if applicable)
- editor, reviser, compiler or translator (if other than the author) *
- publisher *
- place of publication *
- page numbers (when applicable, for example for a chapter in a book)

When referencing material from *periodicals* (or journals) the following order is required:

- author's surname and initials *
- year of publication *
- title of article *
- title of journal or periodical (in italics)*
- title of series (if applicable)
- place of publication (to distinguish different periodicals with the same title)
- volume number *
- issue number (if applicable)
- page number or numbers *

Examples of author–date style

1 Books and similar publications
No author
Computer graphics interfacing 2006, 3rd edn, Modern Technology Corporation, Minneapolis.

One author
Everitt, B.S. 1996, *Making sense of statistics in psychology,* Oxford University Press, Oxford.

Two authors
Shaughnessy, J.J. & Zechmeister, E.B. 1997, *Research methods in psychology*, 4th edn, McGraw-Hill, Singapore.

More than two authors
Hay, I., Bochner, D. & Dungey, C. 2006, *Making the grade,* 3rd edn, Oxford University Press, Melbourne.

Edited work (editor's role not paramount)
The young persons' guide to anarchy 2005, ed. Heinrich A. Stumpendorfer, Hard Core Press, Berlin.

Edited work (editor paramount)
Broom, T.E. (ed.) 1990, *Analytical chemistry: theory and techniques*, M. Dekker, New York.

Chapter in edited book

Shark, J.J. 2003, 'Hormones in fish', in R.L. Whiting & S.S. Trout (eds), *The fishing industries of the South Atlantic,* 5th edn, Seaside Publishing Company, New Hampshire, pp. 31–59.

Book sponsored by an institution or corporation.

Centre for Student Learning 1996, *Methods of learning in science,* ed. J. Davies, HarperCollins, Sydney.

Conference paper (published in conference proceedings)

Smith, F.L. 1996, 'An academic orientation program for commencing students', *Proceedings of the Australasian Science Education Research Association*, University of Canberra, Canberra, pp. 24–32.

Government publication

Australian Bureau of Statistics 1990, *New technology approvals in Australia,* cat. no. 7779.1, ABS, Canberra.

Government report

Department of Education, Science, and Technology 2002, *Higher education at the crossroads: ministerial discussion paper*, DEST, Canberra.

Topic or course materials

School of Engineering, Flinders University 2006, *Project communications course book*, ENG 1001, semester 2, Flinders University, Adelaide, p. 135.

2 Periodicals
Journal (no author)

'New methods of laser detection' 1994, *Laser Technology,* vol. 25, p. 309.

Journal (one author)

Ramsden, P. 1985, 'Student learning research: retrospect and prospect', *Higher Education Research & Development,* vol. 4, no.1, pp. 51–69.

Journal (more than one author)

Boyl, E.A., Duffy, T. & Dunleavy, K. 2003, 'Learning styles and academic outcomes: the validity and utility of Vermunt's Inventory of Learning Styles in British higher education', *British Journal of Educational Psychology,* vol. 73, pp. 267–90.

Journal (no volume or issue number)

Crassinmore, F. 1994, 'A better mouse-trap?', *Creative Technology,* September, pp. 32–5.

Newspaper article

Martyr, H.E. 1994, 'The debate into racism in Australia', *The Australian,* 13 August, p. 14.

Magazine article

Raloff, J. 2005, 'Lead therapy won't help most kids', *Science News*, vol. 159, 12 May, p. 292.

3 Electronic sources
Journal article from a full-text online database

Kember, D., Biggs, J. & Leung, D. 2004, 'Examining the multidimensionality of approaches to learning through the development of a revised version of the LPQ', *British Journal of Educational*

Psychology, vol. 74, no. 2, pp. 261–79, viewed 4 May 2006, <http://proquest.umi.com/pqdlink>.

Article from an electronic journal

Frantzi, K. 2004, 'Human rights education: the United Nations endeavour and the importance of childhood and intelligent sympathy', *International Education Journal,* vol. 5, no. 1, viewed 19 May 2006, <http://www.iej.cjb.net>.

Website with no specified author or date

Flinders University Student Learning Centre, *Article reviews,* viewed 20 December 2006, <http://www.flinders.edu.au/slc/skills.htm>.

CD Rom

The Comedy Store 2002, *The art of comedy,* CD-ROM, Western Publishing, New York.

Electronic mail list or bulletin board

Scrooge, E. <scrooge@dickens.ac.uk> 2006, 'The life and times of a superstar', list server, 1 April, Dickensian Society of London, viewed 5 April, <http://www.dickens.org.uk/home>.

Email

Savonarola, A. 2006, email, 24 January, <Savonarola@santamaria.fir.it>.

Computer program

Pamula, F. & Zeegers, P. 1996, *Writing in the sciences,* Windows 95, Science Software Consortium, Flinders Technologies.

4 Other sources

Standards

Standards Association of Australia 1991, *Australian standard for cauliflowers and cabbages,* AS 1455–1991, Standards Australia, Sydney.

Patent

Tan, I.S. & Arnold, F.F. (US Air Force) 1993, *In situ molecular composites based on rigid-rod polyamides,* US Patent 5 247 057.

Pamphlet

Quit: give smoking away in 5 days 1997, Victorian Smoking and Health Programme booklet, Health Department Victoria, Anti-Cancer Council and National Heart Foundation, Melbourne.

Video, television

The life of the Atlantic salmon 1999, video recording, Canadian Wildlife Services, Vancouver.

Microfiche

Wells, H.G. 1887, microfiche, *The Martian canal industry: an overview,* Australian Science Fiction Council, Canberra.

Musical recording

Waits, T. 1985, cassette recording 61180-4, *Blue valentine,* Asylum Records, Oakland, California.

Legal Case

The State of South Australia v. The Commonwealth (1915) 20 CLR 54.

Reference to Legal Authority

Latham, C.J. & McTieran, J. in *Attorney-General (Vic.) v. The Commonwealth* (1946) 71 CLR at 253-6 and 273-4.

Thesis

Trout, N.A. 1996, *Spectroscopic, stereochemical and reactivity studies in the adamantane ring system*, PhD thesis, Flinders University, Adelaide, Australia.

Unpublished conference paper

Melanti, B.G. 2002, 'Programmers' attitudes toward computer crime: the case in Hong Kong', paper presented to 10th World Congress of Computer Technology, Kathmandu, 16–21 August.

The Vancouver style

One of the variants of the author–date style of referencing, and one commonly used in the medical and physical sciences, is the Vancouver style.

Citations in the text

Instead of citing the author and the date for the in-text citation, this system uses numerals in parentheses; the number in parenthesis links directly to the reference list at the end of the piece of work.

> *Smith (1) has argued that the relative seriousness of the two kinds of statistical error stems from a misunderstanding of the difference between a Type I error and a Type II error. More recently the study by Chan (2) has shown that statistical error using large sample sizes in the biological sciences is not recognised as being serious.*

Some disciplines prefer the use of *superscripts* for the reference in the text, for example:

> *Smith[1] has argued that the relative seriousness of the two kinds of statistical error stems from a misunderstanding of the difference between a Type I error and a Type II error. More recently the study by Chan[2] has shown that statistical error using large sample sizes in the biological sciences is not recognised as being serious.*

Reference list

References should be numbered consecutively in the order in which they are *first* mentioned in the text; they should *not* be listed alphabetically by author or title or put in date order. So the references for the article extract above would look like:

> *1. Smith A. Statistical errors: a case of uncertainty. Ther Med 1999; 7(4): 245–9.*
>
> *2. Chan C.C. Overlooking errors. J Cell Sc 2005; 113(18): 3125–6.*

Note: Do not put parentheses round the numbers for each reference in the reference list.

Examples of Vancouver style of citation of references in the reference list

The following is an abridged version of the Vancouver style of referencing. For a more complete list of possible references please refer to the bibliography or websites at the end of this book.

Books

You need to give details of the authors, title, edition if not the first, place of publication, publisher, and year of publication.

Everitt, B.S. *Making sense of statistics in psychology*. Oxford: Oxford University Press; 1996.

Hay, I., Bochner, D., & Dungey, C. *Making the grade*. 2nd edn. Melbourne: Oxford University Press; 2002.

Edited book

Broom, T.E. (ed.). *Analytical chemistry: theory and techniques*. New York: M Dekker; 1990.

Johnson, C.D. & Taylor, I. (eds). *Recent advances in surgery* vol 23. Edinburgh: Churchill Livingstone; 2000.

Chapter in book

Shark, J.J. 'Hormones in fish'. In R.L. Whiting & S.S. Trout (eds), *Anatomy and physiology of South Atlantic fish species*. 5th edn. New Hampshire: Seaside Publishing Co; 1993. p. 92–101.

Chapter in research monograph

Zeegers, P. & Klinger, C. 'Changes in tertiary science over the last decade'. In C. Bond & P. Bright (eds). *Research & Development in Higher Education: Learning for an Unknown Future*. vol 26. Milperra (NSW): HERDSA Inc; 2003. pp. 630–9.

Conference paper *(published)*

Smith, F.L. 'An academic orientation program for commencing students'. In Proceedings of the Australasian Science Education Research Association, 1996 July 7–9; University of Canberra, Canberra; 2006. pp. 24–6.

Conference paper *(unpublished)*

Smith, T.J. & Jones, H. 'The effect of exogenous opioids on serotonin levels.' Paper presented at the Royal Australasian College of Physicians National Conference, 2005, Sept 12–14, University of Queensland, Australia.

Government publications

Commonwealth Department of Education, Science, and Training. *Higher Education at the Crossroads*. Canberra: DEST; 2002.

Reports

Connought, A. 'Anxiety and mood disorders in young adults' (Australian Health assessment. review; vol.1, no. 6) Canberra: Research on behalf of the MHCA; 2003.

Southampton University Hospitals NHS Trust CHI action plan / Southampton University Hospitals NHS Trust. Southampton: Southampton University Hospitals NHS Trust; 2000.

Dissertation or thesis

Trout, N.A. Spectroscopic, stereochemical, and reactivity studies in the adamantane ring system [dissertation]. Adelaide: Flinders University; 1996.

Dictionary and similar references

Concise Science Dictionary. Oxford: Oxford University Press; 1984. Dictyoptera; p. 199.

Newspaper article

Illing, D. 'Bishop loosens red tape'. The Australian 2006 June 7; p. 35.

Journals

The titles of journals should be *abbreviated* according to the List of Journals Indexed in, for example Index Medicus (Medline), which covers medical science research. List the first six authors followed by et al. A few examples of the different types of journal articles are shown below:

Cancer in South Africa [editorial]. S Afr Med J 1994; 84:15.

Enzensberger, W. & Fischer, P.A. Metronome in Parkinson's disease [letter]. Lancet 1996; 347: 1337.

Clement, J. & De Bock, R. Hematological complications of hantavirus nephropathy (HVN) [abstract]. Kidney Int 1992; 42: 1285.

Ramsden, P. Student learning research: retrospect and prospect. Higher Ed Res & Dev 1985; 4(1): 51–69.

Boyl, E.A., Duffy, T. & Dunleavy, K. 'Learning styles and academic outcomes: the validity and utility of Vermunt's Inventory of Learning Styles'. In *British Higher Education*. Br J Ed Psych 2003; 73: 267–90.

Popplewell, E.J., Innes, V.A., Lloyd-Hughes, S., Jenkins, E.L., Khdir. K. & Bryant, T.N. 'The effect of high-efficiency and standard vacuum-cleaners on mite, cat and dog allergen levels and clinical progress'. Pediatr Allergy Immunol 2000; 11(3): 142–8.

Electronic publications

Department of Health. [No date] Creutzfeldt-Jakob disease: guidance for healthcare workers. [Online] [access 2005 August]. Available from URL http://www.doh.gov.uk/pdfs/cjdguidance.pdf

Department of Health: The Interdepartmental Working Group on Tuberculosis. 1999. The Prevention and Control of Tuberculosis in the United Kingdom [Online] [access 2004 August]. Available from URL http://www.doh.gov.uk/tbguide1.htm.

Journal article in electronic format

Zakrzewska JM. Consumer Views on Management of Trigeminal Neuralgia Headache [serial online] 2001 41(4) 369-376. [access 2005 29 August]. Available from URL http://www.swetsnet.nl/link/access_db?issn=0017-8748

Monograph in electronic format

CDI, clinical dermatology illustrated [monograph on CD-ROM]. Reeves JRT, Maibach H. CMEA Multimedia Group, producers. 2nd edn. Version 2.0. San Diego: CMEA; 1995.

Computer file

The use of computer graphics in diagnostic medicine [computer programme]. Version 5.2. Melbourne: Diagnostic Medical Systems; 2006.

Discussion list

Justice R, [2001 October 5] Alternative medicines for the treatment of asthma [online] [access 2005 October] Available:http//www.jiscmail.ac.uk/lists/ASTHMA.html.

Keeping track of your references

Keeping track of all the reference materials you may use for a major assignment requires you to be highly organised. To assist you in this task, there are several bibliographic programs available, of which the best known is *Endnote*. This program allows you to organise references, create your own reference libraries, reformat references using different referencing styles, do online searches in databases, and insert citations into your text. *Endnote* may be freely available to you through your university library; otherwise it is a program well worth your investment.

Summary

Referencing your sources is an academic skill that you will be expected to master as part of your development as a writer. Each referencing style has certain conventions that may be specific and vary from one discipline to another. Using the appropriate referencing style is part of the broader concept of academic integrity.

12

Writing for the Sciences

How do I know what I think, until I see what I say.

E.M. Forster, English writer (1879–1970)

Key Concepts

- What constitutes good writing?
- Developing and using a writing plan
- Analysing an assignment question
- The concept of themes
- The different sections of a written assignment
- Using data effectively to support an argument
- Using appropriate language and style
- An assignment final checklist

Introduction

In the sciences, like most professional areas, the ability to write well is considered one of the fundamental skills of effective communication. Reports, article reviews, essays, and other written assignments will form an essential component of the assessment for many of your topics. The particular approach to writing used in science-based courses is designated as *scientific writing*. This style may be initially unfamiliar to most students new to university but it is one that you must master. Writing is a skill often undervalued in the sciences and one not often explicitly taught. It not only allows for effective communication, but also gives you an opportunity to develop a number of associated skills, including:

- retrieving, selecting, collating, and analysing relevant resources
- learning to construct a coherent structure and logical argument using these sources

- expressing clearly and concisely your views on scientific research and theories
- learning and understanding content through the writing process

The approach taken in this chapter is deliberately generalist in an effort to make it applicable to a range of writing exercises in sciences. It is designed to help you to develop a writing style that is adaptable to multiple purposes. You may however find that each science discipline (for example biology, psychology, physics, or information technology) may have specific requirements that deviate from the approach presented here. If in doubt, ask for advice from your lecturers, tutors, or from learning advisers. It is important that you develop an individual approach to academic writing but one that incorporates the conventions of the genre. At the end of this book a list of reference books is included which you may consult in order to get a broader perspective of writing in the sciences.

The written communication skills you gain will be of benefit to you not only during your undergraduate course at university but throughout your professional life. What this chapter aims is to give you an approach to writing which you should be able to modify for the specific written tasks asked of you in different topics and beyond.

What constitutes good writing?

Perhaps we can answer this question if we look at the ten key qualities that markers of written assignments at university look for when assessing written work.

1   There is evidence of good planning and structure.
2   The topic in question has been adequately addressed.
3   The material is easily read by using appropriate headings and subheadings.
4   The ability to convince the reader with the arguments presented is demonstrated.
5   Information is well integrated by the use of supporting evidence.
6   Key concepts are elaborated in sufficient depth.
7   Appropriate format, style, and language are used.
8   Appropriate use of referencing is used.
9   There is evidence of original thought.
10   Work is relevant and persuasive.

In order to develop a series of strategies that may assist you to achieve the requirements of good scientific writing as outlined in the ten points above, let us divide the writing task into a simple plan that has essentially four components.

Preparation

1   Analysis of the question.
2   Source materials and referencing.

Writing

3   The plan and structure of the assignment.
4   Language: putting it all together.

In order to tackle any piece of writing it is important that you first decide how you are going to approach the specific task demanded of you. One approach, which is often used but is not necessarily the only one, is to follow a flow chart like the one depicted in figure 12.1. This approach will at least give you a starting point and a general scheme to follow which you may then adapt to your individual needs or style. It is important to develop an approach that works for you.

Figure 12.1   A schematic approach to the process of writing

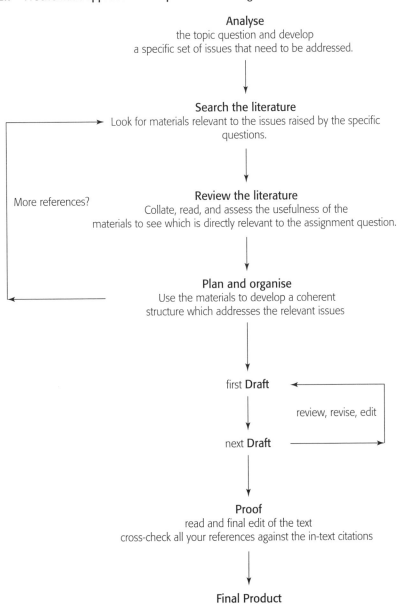

Analyse
the topic question and develop
a specific set of issues that need to be addressed.

Search the literature
Look for materials relevant to the issues raised by the specific
questions.

More references?

Review the literature
Collate, read, and assess the usefulness of the
materials to see which is directly relevant to the assignment question.

Plan and organise
Use the materials to develop a coherent
structure which addresses the relevant issues

first **Draft**

review, revise, edit

next **Draft**

Proof
read and final edit of the text
cross-check all your references against the in-text citations

Final Product

One of the possible shortcomings of this approach is that it gives the impression that writing is a simple linear process with each step following neatly on from the one preceding it. This is clearly not the case for all writing. Writing can be a difficult and often convoluted process and one that requires practice. So let us look at these four key steps.

Analysis of the question

In the eagerness to get started, it is easy to rush into locating and reading lots of literature and to accumulate vast amounts of information, without thinking about what it is that you should actually be looking for, and why. Spend some time thinking about the assignment question. Ask yourself:

- What are the key aspects to be addressed in the question?
- What information do I need to address these aspects?
- How does this question relate to the knowledge in this area of study?
- What is my desired outcome of this piece of writing?
- What is the best way to organise the relevant information into a coherent format?
- Is there a particular format, structure, or style that I must adhere to?

Only when you are clear about the requirements of the writing task and have defined the necessary terms and the parameters are you ready to move to the next step. The key at this stage is to break the assignment question down to a number of smaller questions or issues, each of which should then become one of the headings when it comes to writing.

Example of assignment questions

An assignment question may be very broad or very narrow, it may consist of a complex question dealing with multiple issues, or it could be as simple as a single word. The easiest assignment questions to answer are those that spell out exactly what is required to answer the question. For example:

Explain why hydrogen bonds are important to the structure of DNA.

Other types of questions for an assignment allow you to interpret what may be required and thus you can develop an area related to the broad question that may be of interest to you. Let us take the following topic and develop an assignment question around this topic by asking a number of questions.

AIDS: the 21st century plague?

The following is an example of how this question can be broken down in order to generate a more specific area of discussion. Remember that the questions you want answered should form the basis of the assignment and that there is no defined *correct* answer.

Questions you may want to ask

- What is AIDS?
- What are the effects on different organs of the body? What are the secondary effects? What are the morbidity and mortality rates?
- What is the nature and origin of HIV and how is this related to its mode of action?
- How is HIV spread? What individuals or groups are most susceptible and why? (You may need to consider social or moral issues.)
- Why is HIV so virulent? How endemic is AIDS in the general population, as opposed to specific *at risk* groups?

This trial assignment has used the key word AIDS and it is up to the writer to focus on one possible area and elaborate on that, so you need to set specific questions, the answers to which will then form the focus of the assignment.

So the question could become:

What are the structural features of the AIDS virus that make it so virulent and resistant to effective treatments?

The answer to this particular question would then form the basis of answering the broader question. This should then guide us toward the type of reading material we need to gather from the literature in order to adequately answer the question.

Source materials

The next task is to find literature that will answer your specific enquiries. Rather than repeating information that is available in other parts of this book, please refer to the relevant information on research (Chapter 5) and critical appraisal (Chapter 8).

Planning the structure of the assignment

Having collected sufficient reference material to address the relevant issues raised by the question, your next task is to put that information into some logical order. This *integration* involves making connections between related pieces of information. It is the process of making a big picture from a collection of small pictures, much like the process of making a jigsaw puzzle.

One useful way in which this may be done is to put all your pieces (that is your notes) on a large table or on the floor and move them around till you feel that a coherent picture emerges. It may be that during this process you realise that there are missing links. If this is the case then more research is needed to find those missing pieces of information.

Keep in mind while processing your information why you are doing this. That is, what was the original question you are attempting to answer? Remember that you are trying to construct a logical sequence of other people's information that will answer the overall question in terms of what was done, and how and why it was done. Your *story* must have a beginning, a middle, and an end. One way to do this is to sort the material by themes.

The idea of themes

In essence a piece of writing is a series of interconnected *themes*. The overall theme (hyper-theme) is that suggested by the title. The theme of each paragraph is a particular aspect of the main theme. In turn each paragraph is made up of a series of sentences that have as the sub-theme a specific example or supporting evidence for the idea presented in each paragraph.

The themes and sub-themes are then often used as the *headings* and *subheadings* in the assignment.

The following is an example of the concept of themes using a basic assignment question:

The biological effects of radiation

Let us now suppose that we are to address this question, and in particular the aspect we wish to concentrate on is the effect at a cellular level.

The more specific question (theme) we will attempt to answer is:

The biological effects of radiation: effects on growing and dividing cells

So our assignment can be broken down into a number of smaller interrelated themes and sub-themes as follows:

Example

Theme 1   The nature of radiation

Sub-themes: (a)   Radioactive elements

(b)   Types of radiation

Theme 2   Sources of radiation in the environment

Theme 3   Gross biological effects of exposure to radiation

Theme 4   Effects of radiation at a cellular level

Sub-themes: (a)   Effects on cell division

(b)   Chromosomal effects

(c)   Effects on rapidly dividing tissue

(d)   Adverse effects on cell growth

Themes 1 to 3 are really background information about radiation and its effects on biological systems. The main body of the assignment is theme 4 where we look specifically at the cellular level and address four different aspects (sub-themes). Each of these is then addressed using specific examples from the literature.

Sections of the written assignment

Let's consider a second assignment topic as another example of the general topic of radiation.

Environmental effects of the storage of nuclear wastes in Central Australia.

The following is an example of a suitable structure for your writing and is divided into four main parts. The purpose and the content of each part are briefly outlined.

Introduction

The introduction sets the *context* for the question and should contain the following:

The basic theme

The first step is a general statement about the broad scientific issues to be discussed. An example may be:

The role of nuclear fuels as an alternative to fossil fuels as a source of energy for the next century is an issue of considerable international debate. Some of the issues of concern include …

Background information

(i) General background includes a statement about the broad scientific ideas and how they are related to the main theme. For the question above the broad question may be any of the following:

- Future energy requirements and sources
- Nuclear waste disposal technologies
- Current examples

(ii) Specific background is the part of the introduction where you need to cite recent works relevant to the question. In this section you need to narrow down the broad outline from above and concentrate on a particular aspect.

Example

The long-term storage of spent nuclear fuels is a question with important environmental, social, moral, and political ramifications. The recent findings of Smith (2003) and Nurk (2005) suggest that current technology is not sufficiently advanced to allow for the long-term storage of radioactive wastes in a safe way, though various approaches have been suggested.

Tracey (2005) has advocated the use of a synthetic rock, synroc, to trap nuclear wastes, which can then be stored in deep mine shafts. Macey (2006) found that synroc was not completely safe under certain environmental conditions and suggested that its use be suspended until further investigations have been carried out.

Aims

This consists of questions that are to be specifically addressed in this assignment and the relevance to the broad issues. You will need to spell out clearly the specific aims of your assignment.

Example

This essay will address the issue of nuclear waste disposal, specifically the storage of medium level radioactive materials in Central Australia. Firstly, what are the possible modes of leakage of radiation into groundwater, air, and topsoil? Secondly, what are the possible effects that this storage may have on the fauna and flora of the Cooper Basin?

Body of the text

There are a number of things that you should keep in mind when writing the main body of the assignment. The questions from the aims form the framework of the body of the assignment. It is suggested that you subdivide the information into a number of headings and subheadings with a logical progression of ideas. Do not fall into the trap of trying to cover too much ground as this will result in you writing about many issues in a shallow and superficial way. It is preferable to cover a small number of issues in depth by recognising areas of commonality or difference.

Each paragraph looks at a single issue and uses *specific evidence* to support the arguments that you present, by referring to and using data in the form of tables, diagrams, figures, and graphs from your reference material.

After presenting the evidence for your argument, express your opinion on the meaning and implications of this evidence in the context of your assignment question. This will indicate some original thought on your part.

Link together each paragraph and the individual items of information presented to give a summary at the end of each section.

Summary and conclusions

At the end of the assignment comes the most important part of the whole assignment, and that is the conclusions. In order to complete this section effectively you will need to *draw together* the ideas and evidence presented in such a way that it answers the basic questions with which you started in the introduction.

Writing a summary and conclusions

Integrate

First, you need to present a summary of the information used in the body of the assignment, by integrating all the key pieces of evidence.

Conclude

Second, from this integrated summary you need to make a general conclusion. This is an interpretation of the meaning of the evidence presented and its relation to the original questions asked in terms of the specific aims.

Speculate

Finally, you need to relate the overall conclusions to the broader scientific question behind the assignment topic and possibly highlight the need for further study or suggest future questions.

References

At the end of the assignment you list all of the sources of information cited in the text, including experimental data. The system of referencing you use must be consistent and conform to one of the recognised referencing styles, such as author–date (Harvard) or Vancouver, or some other system your discipline dictates.

Using graphs, tables, and figures

Often in written assignments in the sciences specific information is used to support an argument. This information may be presented in the form of data, graphs, or tables. In order to do this correctly and effectively you will need to do two things:

1 Give all the information required to interpret the diagram.

2 Use the diagram to support a statement.

Imagine figure 12.2 has been used in an assignment that investigated the effectiveness of an antibacterial agent (Chloramphenicol) to destroy several known varieties of bacteria.

Figure 12.2   The viability of *Escherichia coli, Staphylococcus aureus,* and Cyanobacteria after exposure to Chloramphenicol

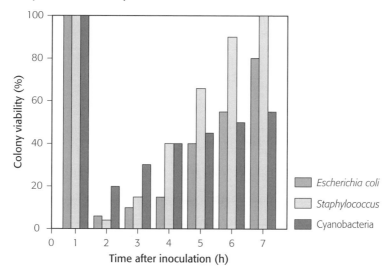

This graph is only useful when it is *used* to support a statement. In order to do this effectively you must point out to the reader exactly why the data are significant. The graph can be used by referring to the following information about it:

- No effect was observable for each type of bacteria after 1 hour.
- There was a rapid decline in bacterial numbers after 2 hours.
- In each case there was a recovery in bacterial numbers over the next 5 hours.
- There is a difference in the type of response between the three bacteria tested. (This will need further elaboration.)

So your assignment may contain a statement such as:

Figure 1 (Smith & Jones 2006) indicates that the antibacterial agent Chloramphenicol is not effective in completely destroying the bacterial colonies tested. It appears to be most effective initially in destroying the Staphylococcus aureus but it was also the most rapid to recover. The Cyanobacteria colony shows the greatest overall reduction in viability in spite of the lowest initial kill rate.

Language: putting it all together

As you are no doubt already aware, the written word comes in many guises, depending on the purpose of its use. The vocabulary, the use of language, the format, and the style used for different written texts is often referred to as the *register*, that is, the version of the written word as applied to a particular context. This comes about due to such considerations as the:

- *topic* being written about
- *audience* to whom the writing is directed
- *purpose* of the writing
- *nature* of the information.

The scientific world uses a register of writing that is different from most other forms of written English. The *scientific register* if often characterised by:

- conciseness
- past tense
- passive verbs
- impersonal (objective) tone
- specific vocabulary and terminology.

Passive rather than active verbs

Let us look at some examples of the type of language used in scientific writing, using passive voice and past tense.

1. A simple sentence:

Active: Science students write assignments.

Passive: Assignments are written by students studying science.

2. A short paragraph:

Active voice:

Madame Curie, the Polish chemist, discovered a new element that has radioactive properties. She called the new element radium. It has the chemical symbol Ra. The element has a mass number of 226 and was first found in 1898. The new element emits radiation of three kinds with a half-life of 1602 years.

Passive voice:

Radium (Ra-226), a radioactive element, was first isolated in 1898 by Madame Curie (Curie & Curie 1899). Ra-226 emits alpha, beta and gamma radiation, and has a half-life of 1602 years.

The second passage has all the essential information, is written in the past tense in a passive tone, is more concise and efficient than the first, and so is the better choice. The subject (theme) of the sentence has been changed so that the focus is on the element radium as an example of a radioactive isotope.

Using too many words

Compare the following two passages.

The female (Homo sapiens), *who is my mate by a ritualised matrimonial and fertility ceremony and who produced you by the process of parturition, has decree, that the consumption of the dried, compressed seeds of the corn plant* (Zea mays) *in association with the lactiferous secretion of the bovine* (Bos bovis) *is a suitable nourishing foodstuff to be consumed as the initial repast of the daylight hours after the period of nighttime inactivity.*

This rather pretentious paragraph could be simply written as:

Your mother says that cornflakes and milk are a good breakfast food.

Though the first statement is more informative and perhaps more *scientific*, there is good argument for the use of the second, as it conveys the same essential information but in a condensed format that is easier to understand. Good scientific writing may lie somewhere between these extremes, such as:

Mothers have argued that cornflakes, made from the dried, compressed seeds of the corn plant, in conjunction with the goodness of milk, are a good source of nourishment for breakfast.

Connecting phrases

It is not enough to have your assignment as a collection of results and findings of other people's work. The next step is to link these ideas and to integrate them into a cohesive argument. One simple method of doing this is the use of words and phrases that connect pieces of information.

Examples

- Similar results were found by …
- In contrast to the findings of …
- This supports the study …
- However …
- These results complement …
- It was also observed by …
- This compares with …
- It was shown that …

The most difficult aspect finally is to relate the information and evidence presented to the questions that form the aims of the assignment. This involves your direct input as *you* must assess the worth of the information and integrate *your* ideas into the arguments presented.

This can be achieved by the use of certain qualifying phrases:

- The evidence presented supports …
- It has been established that …
- The results of the study by Smith (1990) would support the proposal that …
- The rat studies of King (1991) contradict the work of Queen (1987) …
- The evidence of Smith (1989) is the more convincing because …

A proposed structure for scientific writing

Figure 12.3 is a schematic representation of the writing process and the structure of a piece of writing. Note particularly the structure of the introduction, in going from the broad to the specific, and the reverse of this for the summary and conclusion.

The important thing to note about the scheme presented above is that it has a physical structure. We have attempted to show that the introduction and conclusion are mirror images, in that the introduction starts off broadly and narrows to specific aims, whereas the conclusion follows the opposite pattern.

Figure 12.3   A proposed structure for a written assignment

Introduction

The question of the assignment is put into a broad scientific context.

Relevant general background information is presented.

Specific background information
using recent reference material.

Specific aims of this
work.

Body of argument

The questions raised in the aims should be the headings for the main
part of the text.

Each paragraph should focus on one aspect of each part of the question and
on specific information.

The evidence should be presented in a logical progression with data to
support the line of the argument presented.

At the end of each subsection, summarise the evidence presented so far
and offer a brief conclusion as the meaning of the evidence.

Summary and Conclusion

Summarise the main points presented.

What conclusions can be drawn from the evidence?

Do these conclusions address the questions from the aims?

Has the argument presented added to our knowledge in this field of study?

In order to further our knowledge of this field and its relation to the broader scientific issues
what further studies need to be carried out?

Final checklist

The following is offered as a possible checklist for you to consult before you hand in your
work for assessment.

Presentation

☐   appropriate style, format, word limit

☐   title page including your information (name, date, topic)

☐ use of headings and subheadings

☐ correct spelling, grammar, language

☐ appropriate use of references in text

☐ reference list

Introduction

☐ appropriate background information, from general to specific

☐ topic defined in terms of scientific relevance

☐ specific aims clearly defined

Discussion

☐ structure based on the questions to be addressed

☐ evidence of understanding of reading material

☐ logical progression of content and argument

☐ appropriate use of diagrams, tables, figures, and data to support arguments

☐ appropriate breadth and depth devoted to issues discussed

☐ evidence of original thought based on evidence presented

☐ clear and concise summaries

Conclusions

☐ summary of the main ideas presented

☐ highlighting of significant information or lack of information

☐ conclusions based on the evidence presented

☐ final conclusions which address the aims of the assignment

☐ suggestion of further studies

Summary

There is no universally accepted form of scientific writing; it varies from one discipline to another and may take a different form in a textbook, a journal article, or an electronic source. You need to make your writing appropriate to the task at hand. This can still imply that you use your own style and words. All scientific writing requires the use of terminology specific to the subject area but you should not use terminology that you do not fully understand. If you are unsure of the exact meaning of terminology, a word, or a phrase then it is best to check a subject dictionary or reference book, or alternatively find a more appropriate term. Ultimately, the person assessing your work wants to see if you have understood the material you have read and that you can draw together pieces of information from different sources into a coherent piece of writing and to express your opinion on a scientific issue.

13

Writing a Report

The solution is always in front of us Watson, if we know how to look for it.

Sherlock Holmes in *The Hound of the Baskervilles* by Arthur Conan Doyle (1859–1930)

Key Concepts

- The functions of a report
- Sections of a report
- The essentials for each section
- Checklist for report writing

Introduction

The management of a business, government department, or private corporation regularly needs access to objective, factual information in order to address problem areas and make recommendations for future planning. A report is usually a commissioned work with specific questions or issues to be addressed and certain criteria used to set limitations. The information presented in such reports may come from a variety of sources including original research, personal experiences, library research, literature and archival reviews, interviews, statistical analysis of data, or a compilation of any or all of these. This information needs to be presented in a format appropriate to the task. This is commonly done using the following schematic structure:

- What is the nature of the problem or issue?
- What are the terms of reference? Who, what, when, why?
- How were the results, data, or information obtained?

- What are the key findings?
- What conclusions can be drawn from these findings?
- What recommendations can be made for future action?

Sections of a report

In order to address each of the aspects raised above, most reports have a highly structured format in terms of the way in which they are written. Most reports will contain a number of different sections, as shown below, with each section having a specific purpose. Not all reports will contain each of the sections listed, so those sections which may be optional are designated with #.

Title Page

Acknowledgments #

Abstract or executive summary

Table of Contents

List of figures #

Introduction

Materials and methods #

Main findings

Discussion

Conclusions

Recommendations

References

Appendixes #

Let us look at a fictitious report, the title of which could be:

Storage of nuclear waste at Maralinga: an environmental impact study

Title page and inside cover

The title page of the report normally just consists of the title of the report and the name of the authors. The inside cover contains more specific information including the following (figure 13.1):

- What is the specific issue or question being asked?
- For whom is the report prepared?
- Author's name and affiliation
- Date.

Figure 13.1   Example of an inside cover of a report

Storage of Nuclear Waste at Maralinga:
An Environmental Impact Study

The viability of the storage of low-level and medium-level nuclear waste at the
former nuclear test-site of Maralinga in the far north of South Australia.

Dr X.T. Remelycareful
Department of Environmental Studies
Flounders University
Adelaide SA

A report prepared for
The Honorable Alexander Whatsisname, MP
Minister for Environment and Planning
Government of South Australia

Adelaide

August 2006

Note, however, that when you hand in a report for assessment in a study topic you
will need to replace this type of cover page with your own details, including your
name and student number, the topic details, date, and any other relevant
information.

Terms of reference

One of the key pieces of information that the reader requires is the terms and conditions
set down for the report. These are not always clearly detailed within the report itself, but
if available should be included. Below is a brief example:

> *This report was commissioned by the Government of the State of South Australia to
> investigate the viability of the long-term storage of low-level and medium-level nuclear
> waste material at the previous nuclear facility of Maralinga. The report must evaluate the
> most effective and cost efficient means and must be tabled to a parliamentary select committee
> chaired by the Minister and scheduled for 21 January 2007. The report will outline the
> facilities needed to be in place to allow the program to commence in January 2010, including
> a cost analysis.*

Table of contents

The table of contents (figure 13.2) must reflect the basic themes under which the report was prepared and be presented using the section or point method:

*   allows for easy reference

*   allows the reader to see at a glance the different subsections

*   page numbers indicate select reading.

Figure 13.2   Example of a contents page of a report

Contents

Executive summary or abstract

An executive summary is an abbreviated synopsis that encapsulates all the main points of the report, and is usually a few hundred words in length, with the following characteristics:

- precise summary of the issues to be investigated
- brief outline of the approach, procedure, and methods
- complete summary of the principal findings
- main conclusions and recommendations.

Introduction

The introduction should set the scene for the report. It should provide the background relevance and context of the report and set the key questions. The introduction should include:

- relevant background to provide the context of the report
- statement of the problems being addressed
- statement of the specific aims of the report
- review of what is currently known and its relationship to the current report
- methods used to address the current issues
- outline of the data and information to be presented
- a statement of any assumptions or limitations.

Materials and methods

This section should give enough specific detail of the methodology employed so that the reader knows precisely what was done and how it was done. Many reports use pre-existing data, and thus a detailed explanation of the materials and methods may not be necessary. If this is not the case, then the following need to be included:

- how the data or information was obtained or provided
- experimental design for data collection
- apparatus, conditions, procedures, subjects, limitations for data collection
- other limitations or factors needed to be taken into consideration.

Main findings

The main text of the report consists of two parts: the results and a discussion of the results. In some reports, these may be presented separately, for example a report relying on generation of data from primary research conducted by the authors of the report. Other reports combine the results of data which may be a collection of data provided by a secondary source (for example the Australian Bureau of Statistics) with the discussion of these results.

1 Results

The results deals with the specific information or data that has been presented and which is used to argue for a particular point of view or support the recommendations. You need to show:

- the nature of the data or information
- logical order for the presentation of the data or information
- tables, graphs, or diagrams where possible
- statistical analysis for summarising data.

2 Discussion

The discussion is the real heart of the report and comprises an analysis and interpretation of the data or information in relation to the specific aim of the report. You need to consider:

- how you interpret the meaning of the data or information
- if the results are significant and precise
- if they add to what is already known
- the results in relation to the purpose of the report
- any irregularities
- if the results compare with previous work
- plausible explanations.

3 Conclusions

The specific conclusions are drawn from the data and their interpretation. That is, what do the results mean in terms of addressing the original problem? The conclusion should include:

- a clear and concise summary of the principal findings
- conclusions that can be justified from the findings
- conclusions that are applicable to the original aims of the report
- if the results are conclusive, and if not, why not
- further studies that could or should be done
- a statement of the author's evaluation of the validity and usefulness of the conclusions.

4 Recommendations

The recommendations made should come directly from the conclusions. That is, a recommendation is really a statement of the following kind:

> *In order to address the issues associated with nuclear waste storage raised in this report, and inclusive of the limitations discussed in the introduction, the following recommendations are proposed …*

The recommendations are usually placed at the end of the report, after the conclusions. However, it is also not uncommon to list the recommendations at the front of the report immediately after the contents page. Some reports place the recommendations at the end of each section discussed in the main body of the report. The recommendations need to be:

- specific
- practical
- relevant to the terms of reference
- based on the conclusions.

References

All sources of information must be cited using a standard referencing format such as the Harvard system.

Appendixes and attachments

At the end of the report it may be worthwhile adding a list of source materials which have been used in compiling the report. The specific details of the appendices do not need to be read by the reader, but are made available should the reader wish to consult them. Some examples of what may be found in appendices or attachments include:

- a list of people, companies, or organisations who put in submissions for the report
- raw data, tables, interview texts, photographs, maps, or any other information relevant to the report
- examples of questionnaires
- statistical analysis
- worked examples of data analysis.

Summary

A report is one example where you can show your ability to outline a specific problem, find and analyse relevant information to address that problem, and construct a concise and succinct piece of writing to provide others with a number of possible recommendations which may solve that problem.

Final checklist

The following is presented as an example of a general checklist for you to use when writing a report.

Format and Presentation

☐ cover (title, name, date) and table of contents

☐ correct style, format, word limit

☐ appropriate use of citations using a recognised referencing system

☐ reference list in alphabetic order

☐ use of section or point system, headings and subheadings

Executive summary

☐ accurate and informative reflection of the total report

☐ statement of the nature of the problem, the information sought, results, and main conclusions

☐ in some reports, an abbreviated form of the recommendations

Introduction

☐ relevant and appropriate background information

☐ issue or problem clearly defined

☐ specific aims, terms of reference, and limitations clearly defined

Results and discussion

☐ Clear structure to presentation of material

☐ Evidence of the appropriate use of relevant materials

☐ Logical and sequential flow of the content

☐ Suitable use of diagrams, tables, figures, or statistics to support the findings

☐ Appropriate breadth and depth devoted to issues discussed

☐ Clear and concise summaries

Conclusions and recommendations

☐ Summary of the main ideas presented

☐ Highlighting of the most significant information

☐ Conclusions based on the evidence presented

☐ Final conclusions that address the aims of the project

☐ Recommendations based on the conclusions

☐ Recommendations that are specific, practical, and able to be implemented

Language and expression

☐ Clear, logical, and concise argument

☐ Appropriate use of language and terminology

☐ Clear structure of sentences and paragraphs leading to a progression of ideas

14

Writing a Laboratory Report

The great tragedy of science ... the slaying of a beautiful hypothesis by ugly facts.

Thomas Huxley, British scientist (1825–95)

Key Concepts

- Essential components of a laboratory report
- Using appropriate language
- Content of the different components

Introduction

Whatever the desired outcomes of any scientific work, one that is common to all such work is the reporting of the observations and results in a clear, logical, and concise manner. The reader must be able to follow exactly what was done, how it was done, and be able to see the rationale behind the observations and the interpretation of the experimental results. This applies whether you are a chemist attempting to synthesise a new drug, a marine biologist observing the hunting behaviour of leopard seals, or a geologist evaluating the mineral content in ore samples.

In some first-year science topics the report of a laboratory, practical, experimental, or field-work session is no more than a printed handout in which spaces are left for you to add your results. This has advantages in that the report is brief, the format is consistent for all students, and it is quick and easy to mark. The main disadvantage is that these reports do not allow for individual expression and they do not give you the experience of writing a more formal laboratory or practical report as will be expected of you in the later years of study.

Essential components of a laboratory report

The following information is designed to give you the essentials of writing a formal laboratory report. Across the sciences there is a consistency in the writing of experimental reports in that they follow a prescribed format and use certain conventions of language, presentation, and style. The basic components of a laboratory report include:

Title

Abstract

Contents

Introduction

Materials and methods

Results

Discussion

Summary and conclusions

Appendices

References

Different study topics may require you to follow a particular format for writing a report but the standard convention is that *all* laboratory reports contain the essential **IMRD** components:

Introduction	What did you set out to do and why?
Methods	How did you do it?
Results	What results did you get?
Discussion	What do the results mean?

The essence of any good laboratory report is that you are able to effectively communicate to the reader exactly what you have done, in such a way that the reader should be able to replicate your work.

A good laboratory report is based on a good laboratory notebook. Write down everything you do and observe during the practical session in your notebook. This will ensure that you know at all times where you are, where you have been, what you have done, and what you are going to do next. Ensure also that you date all of your work and, to make things easy to follow, use appropriate headings and diagrams.

Language and style

A laboratory report is an account of something that you have done, therefore it is most often written in the *past* tense, using the *passive* voice. Convey all the necessary information in as few words as possible so that the reader could repeat your experiment exactly.

Objective language

Use objective rather than subjective language.

> *Objective: Rats were weighed daily over a two-week period.*

> *Subjective: I weighed the rats every day for two weeks with a Salter scale.*

Active or passive voice

Use the passive voice rather than the active voice.

> *Passive: Acetone (10 ml) was added to the mixture drop-wise.*

> *Active: I added 10 ml of acetone, one drop at a time.*

Past or present tense

When describing something that you have already done as part of the experiment, it is a past event so you should use the past tense.

> *Past tense: The temperature of the mixture was recorded after one hour.*

> *Present tense: An increase in temperature occurs in one hour.*

Examples of lab reports

Compare the following two versions of a passage from a laboratory report.

The experiment uses two lots of white rats with ten rats in each group. The rats are allowed to eat as often as they like. One group of rats has rat pellets, some vegetable matter, and water. The other group has only cornflakes and water. The rats will be weighed every day for two weeks to see if there is a difference in their weights. A progress of the weights will be kept and drawn on a graph. At the end of the two weeks the rats were killed and their livers examined under a microscope.

The experimental animal was the white rat in two groups of ten. All were allowed free access to food and water. The experimental group was fed only cornflakes and the control group was fed a variable diet including rat pellets and vegetable matter. Weights were recorded daily for 14 days, after which the animals were sacrificed for liver microscopy.

The second passage has all the essential information, it is written in the past tense in a passive tone, it is more concise and efficient than the first and so is the better choice. Again, if in doubt, ask for guidance from the people who run the laboratories.

Words to keep in mind when writing

Clarity

What you write should be absolutely clear to the reader. Make sure you say what you mean and mean what you say. If you are not just reporting data or observations but making conclusions or interpretations, make sure your logic is clear so that the reader follows your line of thought. Avoid making assumptions, and be precise in your writing so that there is no ambiguity. It is often a good idea to read back to yourself *out loud* what you have written to see if it sounds right.

Is the following statement clear?

Artificial insemination is when the farmer does it to the cow instead of the bull.

What does the pronoun *it* refer to? This passage would read better as:

Artificial insemination is when farmers impregnate cows with semen taken from a bull.

Avoid the use of words with a meaning that could be interpretative, for example words or phrases such as:

considerable, quite, the majority, rather, slow or fast, somewhat, often, a great deal, many, several, some, probable, possible, it is likely.

Correctness

Correctness means correct both in terms of spelling and grammar, but also accurate numerically if you use data, tables, graphs, and diagrams. Have you made any assumptions? Have you interpreted the data or information accurately and correctly? Are your conclusions correct in that they are based on the information from the experiment? Consider the following statement. What is in the boiling water?

After ten minutes in boiling water, I transferred the flask to an ice bath.

This sentence is better written as:

The flask was transferred to an ice bath after ten minutes in boiling water.

Completeness

Make sure all instructions, methods, and procedures are given. If abbreviations are used you must first define them.

Simplicity

All instructions, procedures and arguments presented should be simple to follow so that the reader knows what you are thinking. This applies not only to the style of writing but also to aspects of theory or explanations of observations. Simplicity can often be interpreted as a sign of clarity of thought and understanding. The best writing is often the simplest.

Some statements may be correct but the message gets lost because of a lack of simple English used.

Readability

Structure the report suitably using clear language with a logical progression of ideas. Use appropriate headings, clear handwriting or typed, short sentences, good clear diagrams, and appropriate spacing. Use diagrams or graphs within the text if you can as this can make for easier reading.

Objectivity

A laboratory report can be a statement of facts and observations and should be presented in an unbiased way. However all observations are open to interpretation so you can often express your opinions as to the meaning of the observations or experimental data. This can often be the case when it is necessary to offer an opinion as to why you think perhaps things did not go according to plan. What could be done to rectify this, or change the experiment to reduce the possibility of errors?

Writing a laboratory report

Again, keep in mind that you may not be expected to include all of the following components but they are included here for completeness, and it may be beneficial to think about the function of each in context of the whole report.

Title

The title must convey exactly what was done or the key idea or objective. In most cases what is required is the name of the experiment, as given in the laboratory course book, your name and student number, and the date. An example is:

Example

**The effect of metal ions on the growth rate of
Staphylococcus aureous**
Charlie Greenthumbs (Student No. 123456)
BIOL 3240: Ecophysiology of micro-organisms
20 March 2006

Abstract

An abstract is a synopsis of the total content of the report in terms of the main objectives, the actual findings, and the conclusions. A good abstract should be concise, informative, and give a brief overview of the main findings.

Example

The bacterium *Staphylococcus aureous* was incubated at 35°C in the presence of varying concentrations of metal ions for a period of 24 hours. Reproduction rates were estimated by measuring the number of viable bacterial colonies on blood agar plates. Results indicate that growth rate is directly related to the metal ion concentrations, with each metal ion tested having an optimum concentration, above which the number of viable colonies declined.

Introduction

An introduction is primarily to convey the reasons for doing the experiment or to present the possible expectations or an hypothesis about the outcomes. It also includes some appropriate background information and previous work to put the current work into context. You may also mention any limitations or assumptions. The third part of the introduction is to state clearly the aim of the present work. This may be achieved by a statement such as the following:

Example

Recent work (Smith & Jones 2006) has shown that the reproduction rate of most strains of bacteria is dependent on the concentration of certain metal ions in the growth medium. The aim of this study was to investigate the effect of specific metal ions (Na, K, Mg, Fe, Zn, and Ca) on the reproduction rate of the bacterium *Staphylococcus aureous* using blood agar as a growth medium.

Materials and Methods

This section describes in detail how the experiment was performed and the conditions under which it was performed. You will need to describe the instrumentation and equipment used as well as the techniques used to gather, collate, and analyse the data from the observations made. Do not rewrite verbatim the laboratory manual but, rather, summarise the essential information. Where specific reference is made to a procedure outlined in the laboratory manual it is often sufficient to simply include a statement such as the following:

Example

The number of viable bacteria was measured following the procedure as outlined in detail in Experiment 8, pages 40–2. In essence, colony numbers were counted under 10x magnification, and the area of the colony estimated using the microscope's inbuilt measurement scale.

The important thing to remember is to follow exactly the requirements for each different topic or subject. If you are not absolutely sure what is required, *ask* the laboratory staff or one of your lecturers.

Results

The results section is a description of the observations that you made. Do not present raw data but collate and present the results in a finished format such as tables, graphs, or statistical summaries. Make sure all diagrams are clearly labelled with all information required to interpret the results given. It should not be necessary for the reader to refer to information from other sections of the report in order to interpret diagrams or graphs.

Example

The growth rate of the bacteria was dependent on the presence and concentration of the metal ions in solution. Figure 1 indicates that in the presence of calcium (as Ca^{2+} ions) growth rate increase up to a concentration of 0.05 M, above which there was a decrease in growth rate.

You may need to seek specific guidance as to how the results are required to be presented as different topics may require different ways of presentation. If statistical data are included do not go into detail of how the results were obtained (you can add this in an appendix if required); it is probably sufficient to say something like:

> *Figure 3 shows the growth rate data presented as the mean and one standard deviation for the presence of all the metal ions under test.*

Another example for a significance test:

> *Figure 4 shows the use of a paired t–test (significance at the p = 0.05 level) for growth rates in the presence and absence of metal ions.*

When using diagrams, each one should have an accompanying section detailing what the results are, how they were obtained and their significance to the whole study.

When referring to the results in a particular diagram, do not just write:

> *... the results are shown in figure 1.*

It is better to express the results more precisely:

> *Figure 1 shows the number of viable bacterial colonies against the metal ion concentration for each of the metal ions tested. The most significant aspect of this graph is ...*

Discussion

The first thing that should be done is to discuss whether or not the initial aims of the experiment have been achieved, and should *not* be a rewording of the results. This can only be done by reference to your results and whether they support (or contradict) the initial proposal. The key question is, *what do the results mean?*

Your results are open to various forms of analysis and hence also to interpretation. In the course of a three to six hour laboratory session the results you gather are generally self-explanatory and are usually designed to clearly illustrate the basic principles behind the experiment. If the data that you have collected do not lead to a clear-cut interpretation then say so. You may mention any shortcomings of the experiment or errors in your own experimental procedure.

Summary and conclusions

This is a way of finalising your discussion by linking together the results and their interpretation. Remember that a summary and a conclusion are *not* the same. A summary is restating what has already been said. The conclusion draws together the ideas presented earlier and an interpretation of what they mean.

Briefly restate your main findings. Say why they are important in the context of what you set out to do and in the broader scientific sense. Refer only to your actual results. Do not introduce anything in the conclusion that has not already been mentioned earlier.

Example

The results of this experiment have demonstrated the growth rate of bacteria in the presence of metal ions. In particular, the results indicate that the growth rate is accelerated in the presence of all the metal ions tested, but that each has a limiting effect. In conclusion, it would appear that normal growth rate is dependent on the presence of these ions, but that each can have a toxic effect if the concentration of the ion exceeds a limit, which differs for each ion tested.

References

All sources of information cited in the report must be listed in the references in the appropriate format. Again ask for clarification for each different topic that you are studying, as there are a number of acceptable formats for citing references.

Appendixes

Appendixes are only necessary if you wish to present raw data, an example of methods used to interpret the data, or possibly some other piece of information from the experiment not used previously but which may be helpful in supporting your arguments.

Checklist
Presentation

- □   title of experiment, name of student, date, course
- □   correct style, format, word count
- □   spelling, grammar, headings, and subheadings
- □   appropriate use of references cited in the text, and in the reference list

Introduction

☐ appropriate background information, general and specific

☐ experiment defined in terms of scientific relevance

☐ specific purpose of experiment outlined

☐ limitations and scope defined, terminology explained

Methods

☐ full description of all procedures used

☐ description of equipment, subjects, materials

☐ outline of methods of data analysis

Results

☐ data presented as tables, graphs, figures

☐ ensure all figures have a title, and relevant information

Discussion

☐ clear interpretation of the results and their significance

☐ evidence of understanding of results relevant to the aims

☐ comparison with previous results from other studies, if known

☐ appropriate breadth and depth devoted to issues discussed

☐ logical flow and progression of content

☐ evidence of original thought based on evidence presented

☐ clear and concise summaries

Conclusions

☐ summary of the main results and their interpretation

☐ conclusions based on the evidence presented

☐ final conclusions which address the aims of experiment

Example report

The example that follows is a laboratory report written for a second-year organic chemistry class. It is not necessary for you to understand the content of this example as it is used merely to illustrate the format, style, and language.

The two-phase nitration of phenol
(Experiment No. 20311)
CHEM2500: Organic Chemistry Lab. II

I. M. Notachemist

(Student No. 9912345)

13 April 2006

Abstract

Phenol was nitrated by a procedure using sodium nitrate, dilute sulphuric acid, and an organic solvent, in the presence of a catalyst. The nitration procedure described is a very effective method for nitration, giving up to 70% yield of mononitrated phenols. The procedure is more rapid and with less unwanted oxidation products than traditional nitration procedures.

Introduction

Aromatic compounds are nitrated by the mechanism of electrophilic aromatic substitution. A wide variety of nitrating media have been employed, of which nitric acid in sulphuric acid (HNO_3 / H_2SO_4) is one of the most common (1). One of the major problems associated with the use of strong acids in the nitration of phenols is the formation of unwanted oxidation products. This experiment explores an alternative nitration procedure in which a solution of sodium nitrate in dilute sulphuric acid and diethyl ether act as a two-phase nitrating medium (2). The key to the success of the method is the use of a catalytic quantity of sodium nitrite as initiator of the reaction.

Materials and methods

All chemicals were laboratory grade and used as supplied. Phenol (10.0 g, 94 mmol) in diethyl ether (200 ml) was added to a mixture of $NaNO_3$ (9.5 g, 110 mmol) in 3M H_2SO_4 (200 ml). After the addition of a catalytic amount of $NaNO_2$ (25 mg), the two-phase mixture was stirred mechanically at 20°C for 24 hr. The ether layer was separated and the aqueous phase extracted with diethyl ether (3 x 100 ml). The ether layers were combined, washed with 10% Na_2CO_3 (100 ml), water (2 x 100 ml) and dried over $MgSO_4$. Solvent removal afforded an orange solid (12.0 g). The total product was extracted with hexane (3 x 50 ml) from which the *o*-nitrophenol was isolated, 6.6 g (46%) as bright yellow needles, mp = 44°C compared to a reference value mp = 44–6°C (3), after crystallisation from aqueous ethanol. The residue from the hexane extraction was boiled in HCl/ H_2O solution (100 ml) with charcoal. Crystallisation afforded 3.3 g (23%) of *p*-nitrophenol as pale tan needles, mp = 112° compared to a reference value mp = 114°C (3).

Results and discussion

The procedure as outlined for the nitration of phenol involves an alternative mechanism to the traditional electrophilic aromatic substitution. The sodium nitrite (in the acidic solution) is a source of NO^+. This in turn acts as an electron transfer reagent, which oxidises the aromatic compound to the radical cation and is in turn reduced to NO.

The nitration procedure as described afforded a 69% yield of isolated and purified mononitration products. Of the isolated product obtained, 6.6 g (67%) was the *o*-nitroisomer and 3.3 g (33%) the *p*-isomer. GLC analysis of the reaction mixture indicated that the ortho:para ratio was 1.5. On this basis a higher isolated yield of the *p*-nitroisomer was anticipated. The lower than expected yield may be attributed to the destruction of the product during the work-up procedure which involved boiling in aqueous acid. A third minor product, detected by the GLC analysis, was identified as an oxidative by-product (1,4-benzoquinone) but which was not isolated. The course of the reaction was followed using GLC analysis of the ether phase. Figure 1 shows the relationship between reaction time and the amount of nitrated product. The graph indicates a curvilinear relationship between reaction time and product formation with a linear increase in product formation up to 10 hr, after which there was a decline in the rate of product formation.

FIGURE 1   NITRATION OF PHENOL

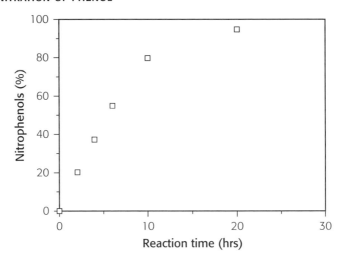

Summary and conclusion

The experiment has demonstrated the nitration of phenol using a two-phase procedure, consisting of an aqueous acidic solution of nitrate and diethyl ether. Catalysis of the reaction rate was achieved by the use of added nitrite. In conclusion, the procedure as described appears to be superior to the conventional mixed acid procedure for the nitration of phenols for the following reasons:

1   The reaction is faster than traditional procedures due to the catalytic action of added NO^+ as $NaNO_2$.
2   The reaction is clean with few tarry oxidative by-products.
3   Using only 10% excess of $NaNO_3$ the reaction was complete with no polynitrated products obtained.
4   High yields of mononitrated products were obtained.

References

1   Vogel, A. 1978, *A textbook of practical organic chemistry,* 4th edn, Longman, New York.
2   Thompson, M.J. & Zeegers, P.J. 1989, The two-phase nitration of phenol, *Tetrahedron,* vol. 45(1), pp. 191–202.
3   *Handbook of chemistry and physics* 1984, 65th edn, CRC Press, Boka Ratou, Florida.

Summary

This chapter has been a brief summary of the format and style of a laboratory report. It is not meant to be a detailed description on writing a laboratory report for a particular branch of science, but a general approach that should suit all sciences. If in doubt about exactly what is expected for your topic, seek help from the people who are in charge of the laboratory classes.

15

Writing an Article Review

I've always been uncomfortable around people who are very certain about their world and their values, no matter how defined. So I find security in pointing out any valid example of contradiction or paradox within their framework of belief.

Ron Cobb, political cartoonist, 1969

Key Concepts

- What is an article review?
- Types of research articles
- Structure of a review

- Critical analysis
- Reviewing an empirical article

Introduction

Almost every day we read, hear, or see advertised a review or critique of a film by Steven Spielberg, a book by Dan Brown, a CD by Wolfmother, or a documentary program on the ABC. In other words, we are exposed to the opinions of a third party about someone else's work. Usually these opinions are meant to encourage us to buy the CD or book, or to see the film or television program. This system of reviewing other people's work also applies to the sciences. In this instance, however, you are not trying to convince others to buy or use a product, but rather you are evaluating another person's research to determine its value to you and other researchers. It may be a report on a new regime for the treatment of diabetes, a more efficient and faster computer chip, or oceanographic data that purports to indicate the current rate of global warming.

Most scientific articles are subjected to a peer review to determine their suitability for publication in research journals. In a similar way, one of the skills that you need to learn and develop is that of critically reviewing the *written* work of others. This may involve evaluating an article, a book, a Web-based journal, the content of a website, or several related publications in various formats. Whatever the medium, you will be assessing the content of the text to see if it is worth reading and recommending.

The review process follows a particular format and is written for a particular reason. You will be assessing the article under review and deciding whether or not it is an example of good research. Consequently, you will be expected to give reasons for your assessment. You need to support or justify your point of view just as you would in an essay.

What is an article review?

An article review is a thoughtful analysis of a piece of writing, usually of a journal article, related to your assignment topic or field of research. A review can be part of an extended annotated bibliography or literature review. The point of the review is to describe the content of the article and its main argument, and to present a discussion about the effectiveness of the article in relation to what it is setting out to show or explain. There are two aspects of an article review you will need to keep in mind:

- An article review is not just a description of the content of the article. It requires an evaluation. To this extent, an article review tends to have a *critical* component.

- An article review is written for a *purpose*. It is important to bear in mind why you are writing the review. Who is the audience and what is the aim behind the review? What are your lecturer's expectations? The purpose will affect how detailed your response is, what areas you need to address, and how critical it needs to be.

Research article types

There are two main categories of research article: empirical articles and conceptual articles. You will need to determine which one you are dealing with as this will impact on what your review critiques.

- An *empirical* article supports its claims by using evidence gathered from experimental data, statistical analysis, field studies, questionnaires, or case studies (primary research data). It will outline its methodology and include a description of how the data was collected, analysed, and interpreted.

- A *conceptual* article supports its claims using logic, theoretical or linguistic analysis, and persuasive reasoning. Although a conceptual article may use empirical evidence to back up its claims, it will only report the findings rather than how the data were collected (secondary research).

Both types of article may use examples to illustrate relevant points. Examples do not, in themselves, count as empirical data. Neither does the use of examples count as evidence of argument; examples merely offer possible support for an argument.

A review of an empirical article will need to include an evaluation of the data presented, the research methods used, and the validity of the interpretation.

Simple structure

Article reviews tend to follow certain conventions. They have a particular format. In the first instance, they should always contain the following elements, in the order listed below.

- title, for example *A Critique of Grant Gillett's Brain Bisection and Personal Identity*
- bibliographic information about the article—author, journal, volume, page numbers (this appears at the top of the review under the title)
- a brief abstract or summary of the essential content (what the article is about), the author's main ideas (what new ideas are being argued for or presented), and a summary of the evidence used to support the claim
- an evaluation of the work presented—its contribution to the field, its limitations or problems, supported by selected examples from the article.

Additional elements

Depending on the purpose of the review, the depth of analysis required, and whether it is a conceptual or empirical article, you may need to include some of the following:

- a brief discussion of the article's aim and how it is developed
- a brief description outlining the article's context
- a discussion of related works in the field and the similarities and differences (relevance)
- a critique of the work, including an assessment of the quality and relevance of the data, the way the data are presented, the overall structure of the paper, and the validity of the conclusions
- an outline of the methodology or theoretic framework used by the author
- an analysis of logical consistency, coherence, and depth of understanding of the issues (conceptual).

Example of an abstract

G.E.R. Lloyd, Adversaries and authorities: investigations into ancient Greek and Chinese science, 1996, Cambridge University Press, Cambridge

Lloyd's aim in this book is to compare ancient Greek and Chinese science to see, first, whether there are genuine differences that can be generalised and, second, to explain why those differences (if they can be shown to exist) emerged. Despite his reservations about overgeneralisation, Lloyd establishes a strong case for differences between the two cultures. These differences, according to Lloyd, are genuine rather than superficial. The reasons for the differences, he argues, are largely sociopolitical. He supports this claim by providing evidence to show that the theoretical concepts underpinning scientific studies in both cultures reflect general political concepts and ideals that were specific to each culture.

This abstract clearly summarises Lloyd's intentions, his main conclusion, and the supporting evidence he uses, but it is *not* an evaluation of the book. It is merely a summary.

Critical analysis

An article review cannot just be descriptive. While it is essential to provide a good summary of the article, letting the reader know what the article is about and what it is claiming, you also need to assess its worth.

One of the easiest ways of critiquing or evaluating a piece of writing is to ask yourself questions about the text. The questions should reflect what you want to assess or evaluate. The answers will provide a detailed framework that can be used to write an in-depth analysis and evaluation of the article. Some suggestions for the kinds of questions to ask are listed below. These are relevant to all articles.

- Is the article appropriate for its target audience?
- Does the article build on prior research?
- Does the article reflect a good knowledge of previous literature in the field?
- Do the authors identify the problem or issue clearly and explain its relevance?
- Did the authors choose the best research method and approach? Was it executed properly?
- Were the methodology, findings, and reasons for their conclusions logically and clearly explained?
- Do the authors make appropriate comparisons to similar events, cases, or occurrences?
- Are the ideas really new or do the authors merely repackage old ideas with new names?
- Were there adequate and appropriate examples and illustrations?
- Do the authors discuss everything they promise in the abstract, introduction, and outline?
- Does the article make a contribution to its field? If not, in what way should it have made a contribution and why didn't it?
- What are the articles strengths and weaknesses?
- What are its limitations or boundaries?
- Did it discuss all the important aspects in its domain thoroughly?
- Overall, how complete and thorough a job did the authors do? Did they justify their conclusions adequately? Did they provide enough background information to make their work comprehensible?
- How confident are you in the article's results? Is it convincing?

Example of a short critique

Peter Breggin, Toxic psychiatry: why therapy, empathy, and love must replace the drugs, electroshock, and biochemical theories of the new psychiatry, 1991, St Martins Press, NY.

Peter Breggin MD is a practising psychiatrist and is known for his opposition to the use of modern therapeutic approaches in the psychiatric setting. This book is his definitive argument against the current popular treatment regimes for psychiatric disorders, which focus on the use of drugs, surgery, and electroshock therapy (EST). Breggin's style of writing is informative and easy to read. The various chapters were broken down into a number of short subtitled paragraphs all of which made the book easy to follow. It is thought-provoking and informative.

Breggin's perspective is sound, and appears to be well-researched and presented, but not enough direct comparisons were supplied. For example, the author did not evaluate the counter-arguments. Breggin describes in great detail his opinions regarding the potential physical and psychological damage mediated through the use of various drugs. However, little space is devoted to an evaluation of the established view in regards to the positive outcomes associated with the prescription of various psychiatric drugs, the use of surgery, and EST as a treatment regime for specific psychiatric disorders. In addition, the section discussing alternative therapies was somewhat limited in the number and types of therapies discussed, and in informing the reader of their limitations. The appendix listed various addresses of self-help organisations but could have been more detailed regarding the contact details of alternative therapy organisations.

Despite these shortcomings, this book presents an interesting though limited argument for alternative therapies in an easily understandable way. Its main failing is in the lack of a rigorous evaluation of current procedures.

Critiquing an experimental or empirical article

While the critical analysis outlined above applies generally to all articles, empirical or experimental articles make up the bulk of the scientific literature and contain sections that do not appear in other kinds of articles. These sections are where the authors stipulate their research methods and findings: how they conducted the experiments, how they analysed the results, and what they found. These need to be appraised separately and according to their merit.

One of the main difficulties you will face when you first start to read and analyse scientific articles is the technical terminology, the jargon of different disciplines, and the level of complexity of the experimental techniques that may be used. Unfortunately, there is no short way around this, so with practice comes familiarity. As you learn more about your particular field of science, you will learn what constitutes good experimental procedure and what counts as enough evidence to make inferences. This knowledge will enable you to assess how well someone has conducted their experiments, whether their findings are reliable, and how thoroughly they have interpreted the results.

A template for reviewing empirical articles

Most scientific articles present 'hard data' from experiments and observations to support a particular claim. The following provides a guide or template for the questions you need to ask to assess the value of an empirical article in terms of the validity of the experimental procedures, research methods, data interpretation, and resulting claims.

Sections of an article

Introduction

- What specific scientific problem or issue is being addressed?
- What is the overall premise or hypothesis to be tested?
- How is the study relevant to the specific issue and the broader scientific context in which it occurs?

Experimental

- What experiments or observations have been performed?
- What specific methods were used and why were these chosen?
- Under what conditions, limitations, and restrictions were the experiments performed?
- How were the data collected, collated, and analysed?
- Were allowances made for potential sources of errors?

Results and discussion

- What were the main findings of the experiments or observations?
- Are the results presented in tables, diagrams, and graphs that are clear and self-explanatory?
- Are statistics presented? If so, are the statistical interpretations accurate and are they based on the experimental results?
- What conclusions have the authors drawn from the results and their interpretation?
- Are these interpretations warranted based only on the results or have certain assumptions been made?

Conclusions

- Do the experimental results and their interpretations and conclusions answer the specific aim of the study?
- Has the study contributed to an increase in understanding in the area of study?

Once you have worked out what the article is about, how it has supported its claim, what its strengths and weaknesses are, and assessed its value, you can begin to write the review. Use the template as a guide. It will provide you with most of the points you need for your review. Remember to make it clear when you are talking about what the authors say in the article and when you are referring to your own ideas or knowledge about the topic. For example:

Smith puts forward the idea that ...

Smith's main proposal is that ...

He argues that ...

While Smith's data supports his claims, evidence from Jones (2006) ...

Other research by Chan (2005) does not support ...

For more information about the writing process refer to Chapter 13.

Summary

Remember, a review puts forward *your* evaluation of an article. This means you are now presenting an argument. Anything you say about the article must be well argued and supported by evidence. You may need to quote from the article and refer to any other sources used. You will also need to justify your claims. This can be done by answering the following questions:

* Is it a good article? Why?
* How could it be improved?

This is the most important feature of the review because you are generating *reasons* for your perspective or point of view. If you follow the above format, your review will be informative, well-argued, and will have a logical structure.

16

Presentation Skills: Talks and Posters

'Don't grunt,' said Alice, *'that's not a proper way of expressing yourself.'*

Alice's Adventures in Wonderland by Lewis Carroll (1832–98)

Key Concepts

- Why oral presentations may fall flat
- The two key elements to oral presentations
- Structuring an oral presentation
- Using flow charts
- Using the right language
- Presenting a poster
- The structure and content of a poster

Introduction

When marketing companies survey people on all manner of topics, one of the most consistent fears expressed by those interviewed is their fear of public speaking. This fear that we are the centre of attention and that all eyes are upon us seems to be universal. There are a number of stories about famous actors, those whose job it is to perform in front of others, who get so nervous before a performance that they are physically sick. Performance anxiety is so well known that it is often the subject of jokes, as illustrated in a scene from the 1965 British comedy *Carry on Cleo,* when the character of Julius Caesar, played by Kenneth Williams, is about to address the Roman senate and utters the line: 'Infamy, infamy, they've all got it in for me!'

University students are no exception to the fear of public speaking, in that the thought of having to give an oral presentation fills most students with absolute terror.

This chapter will give you some strategies which may help to improve the quality of your presentations and reduce the anxiety associated with them. It must be kept in mind that this is intended only as a brief introduction to the much larger topic of public speaking.

Oral presentations

Oral communication is a skill that requires techniques that are different from those required in written communication. Like all skills, public speaking is a learned skill that must be practised in order to be able to effectively communicate complex scientific ideas in a clear, logical, and succinct manner. No doubt you have attended lectures, talks, or presentations, or at least seen people present programs on television, for example on the evening news. It may be difficult to pin down exactly what made the presentation a good one, so it is often easier to identify those characteristics of a presentation that you thought were not so good. Let us look at some of the qualities that differentiate a good presentation from a poor one (figure 16.1).

Figure 16.1   Differences between a poor and a good presentation

A poor presentation	A good presentation
Poor timing or structure	A logical structured sequence which uses the time available
Speaking too quickly, too softly, or mumbling	A clear voice raised just slightly above a normal speaking level
Reading directly from notes	Use of notes as prompts
Information overload: too much, or too complex	Saying only what is needed to get the point across
Inappropriate or poor use of audiovisual aids	Visual material used to highlight or summarise important points
Lack of confidence, nervous fidgeting	Some nervousness, but awareness of your body language
Not speaking to the audience	Looking at the audience, making eye contact

The difference between a good presentation and a poor one is not always as obvious as the distinctions shown above, but if you are aware of some of these key differences it will help you improve your presentations.

The single most important asset for giving a good presentation is self-confidence, and this comes from two key skills, *preparation* and *delivery*.

Preparation

The first step is to be prepared. To assist you in this, the following is a series of points that you should consider so that you are as well prepared as possible.

The audience

- To whom are you speaking—peers, lecturers, the general public?
- What is the audience's level of knowledge of your topic?
- What type of language is appropriate for the audience?
- What will the audience expect of the presentation?

Expectations of the talk

- Why are you giving this talk?
- Are you there to inform, persuade, or entertain?
- How will your presentation be assessed?

The physical constraints

- What are the physical aspects of the location—room size, acoustics, lighting?
- How large is the audience and where are they situated in the room?
- How long will you have to speak?
- How many other speakers are there?
- What is available in terms of audiovisual equipment?
- Can you be easily seen and heard from all areas of the room?

Preparing the presentation

- Have a good title.
- Write down the central theme of the talk in the middle of a blank sheet of paper.
- What other ideas spring from this central theme, and what assumptions have you made about the audience's knowledge of this subject?
- What is the intent of the talk?
- Arrange the content of the talk into:
  - introduction
  - main content
  - summary and conclusions.
- Prepare a suitable set of notes.
- Write prompts on cards or sheets of paper.
- Decide what is most appropriate for audiovisual aids.
- Practice the talk and time yourself.
- Memorise key aspects, such as the opening statement and summary.
- Develop a suitable ending, for example pose a question or propose a particular course of action.

The content of the talk

Introduction

- Begin with a strong statement, some important facts, or a leading question.
- Tell the audience what you are going to talk about.
- Present the context of the talk by giving the relevant background information.
- Emphasise the key questions which need to be addressed.

Main body

Develop the main ideas of the talk by having a clear structure or pattern of its organisation (figure 16.2).

Figure 16.2   Organisation of a talk

problem ⟶ solution

question ⟶ answer

cause ⟶ effect

simple ⟶ complex

Structuring a talk

There are several ways to structure the content of a talk:

- **Chronological.** Points are presented in a time sequence.
- **Spatial.** Points are presented as they relate to each other, for example the big picture leads to fine detail.
- **Cause and effect.** Points flow from a problem, question, or observation.
- **Topical.** Points are arranged in perceived order of importance or relevance to other issues.

Summary and conclusions

- Summarise what you have already said: 'In summary ...'
- Evaluate the importance of the information.
- Draw conclusions from the information.
- Suggest a next step, course of action, or questions for further research.

Presenting a literature review

Many of the talks at undergraduate level involve giving a summary of a scientific article in the recent literature, or discussing a topic presented during a lecture. Generally these talks are quite short and range from a five minute summation to a 30 minute overview.

When presenting a review of a research journal article there are five basic questions that you need to keep in mind.

- What was done?
- How was it done?
- Why was it done?
- What results were obtained?
- What do the results mean?

Flow charts

When you make an oral presentation you most often deal with a specific topic or questions of interest. Figure 16.3 shows a flow chart for the presentation of a research talk.

Figure 16.3   Flow chart of a research talk

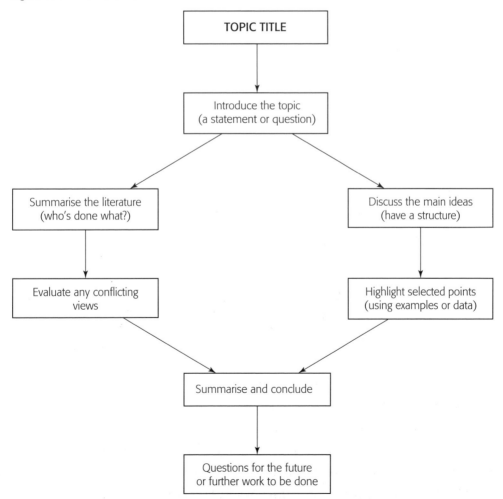

Delivery

Nervousness before your presentation is not only normal but also to be expected. Take a few deep breaths. Focus your mind on the task at hand. See yourself as a successful presenter, perhaps modelled on someone you know is a good presenter. Have your notes and audiovisual aids ready. Walk to the podium or front of the room slowly and purposefully. Take a second or two to gather yourself and the material. Start with a well-rehearsed opening statement or question.

Credibility

Your believability or credibility is based on three things:
*   Verbals (7%): the words you use
*   Vocals (38%): how you say the words
*   Visuals (55%): what you do when you speak.

Verbals and vocals

*   Use language appropriate to the audience.
*   Avoid too much jargon and vernacular that may not be familiar to your audience.
*   Show enthusiasm.
*   Speak clearly and use correct pronunciation of words.
*   Vary the speed, volume, and tone of your voice.
*   Speak to the whole audience, not just the front few rows.
*   Use your voice to emphasise the important points.
*   Do not read directly from your notes.

Visuals

*   Maintain eye contact with the audience.
*   Use gestures to emphasise important points, for example your hands or a pointer.
*   Avoid the use of 'non-words' such as err, uhm, ah.
*   Look relaxed and confident, even though you may not be.
*   Use some facial expressions, but remember you are not Jim Carrey.
*   Maintain a comfortable and relaxed position.
*   Move about occasionally, but not constantly: you are not a lion in a cage.
*   Be aware of any distracting mannerisms, such as playing with keys in your pocket.
*   Use your audiovisual aids for effect, rather than trying to show everything.
*   Maintain a sense of the occasion: this is not a life or death situation.

Visual aids

Visual aids are used to increase the impact of your talk. They should *add* to the oral presentation, not distract from it. Many presentations today use computers with programs such as *Powerpoint*, part of the MS Office package. If you are unfamiliar with this program, make sure you try it out before the presentation. Do not overdo the use of the visual effects within *Powerpoint*, as this may detract from the content of the talk, and keep in mind that the main purpose of visual aids is to (figure 16.4):

* introduce and recapitulate
* emphasise the main points
* show a sequence, such as a flow chart
* present complex information, data, or ideas
* maintain the interest of the audience
* present statistical data
* summarise information.

Figure 16.4   Audiovisual presentation

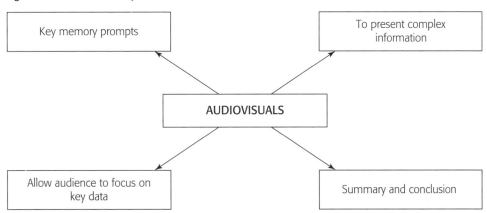

Poster presentations

A poster can be used to present information in a brief, rapid, and visually interesting way. It is often used as an alternative to an oral presentation and is a popular means of presentation at scientific conferences. A poster requires you to be able to express sometimes complex ideas with brevity, clarity, and accuracy in order to stimulate further discussion or debate.

Five basic principles of poster presentation

* **Attention-getting.** You need to attract the attention of your audience in order to make them want to read the poster.
* **Brevity.** Say what you have to in the most efficient manner.

- **Coherence.** A poster should present information in a stand-alone manner. Each statement should be presented in a logical progression without requiring further information in order to make sense.
- **Simplicity.** Keep it simple and focused.
- **Evidence.** Use suitable reference material to state your case.

(Source: Hay, I., Bochner, D. & Dungey, C. 2002, *Making the Grade,* 2nd edn, Oxford University Press, Melbourne, p. 241)

Presentation

The key aspect of the presentation is that the content should be thoughtfully presented in a visually effective way and that there is a logical progression of ideas. The overall structure does not have to be a linear one but it must be simple and obvious. Also remember the saying that a picture is worth a thousand words, hence use graphics effectively.

There are some sections which are considered essential to all posters:

- Title
- Introduction
- The main text
- Conclusions
- References

Your poster layout is really a question of how best to present your results and the data to support them. Figures 16.5 and 16.6 offer two examples.

Figure 16.5   Volcanic emissions poster

Volcanic emissions and global warming Julius Marks, Department of Geomorphology, Curtin University, Australia		
Summary Give an overview of the project in terms of aims, main findings, and the conclusions drawn.	**Aims** What did the project set out to do?	**Results** Describe the outcomes of the project using graphs, tables or other figures. Do not discuss the meaning of the results here.
Methods Outline the specific methods used giving all experimental details, analysis, and statistical procedures, and why these procedures were used.	**Discussion** Explain the meaning of the results and highlight those that are particularly important.	You may need to be selective and choose those results that best illustrate the aims.
	References Cite only those references which are integral to the poster.	**Conclusion** Summarise, conclude, and speculate about the meanings of the main findings.

Figure 16.6   Antiviral extracts

Antiviral extracts from Pacific marine jellyfish Stan Laurels, Department of Marine Science, James Cook University, Australia		
Aims and objectives	Abstract	Discussion
Methods	Conclusions	
	References	

Typeface

You should be able to clearly read the poster from a distance of approximately two metres. Figure 16.7 shows recommended sizes of type to be used for different sections, using a common font such as Times New Roman.

Figure 16.7   Suitable font sizes for a poster in Times New Roman

Main title	Bold	96–180 point
Authors and affiliations	Bold	36–72 point
Main headings	Bold	48–84 point
Section headings	Bold	24–36 point
Text	Regular	14–18 point
Figure captions	Bold	14–18 point

Content

You must be guided as far as the content is concerned by the requirements defined by the subject for which the poster assignment has been set. This may include:

- word limit
- size of the poster
- general structural features
- use of graphs, figures, and tables.

Language

The language you use in a poster is the same as you would use for any written assignment in science: past tense, passive voice, impersonal style, and appropriate vocabulary. Do not make your language short and jerky in an effort to stay within the word limit, but use shorthand methods, such as bullet points or numbered lists, to present related materials.

Summary

The ability to present complex and detailed information, either verbally or through pictorial means, is one of the many skills you will learn as part of your tertiary education. The first crucial step will always be how well-prepared you are. Once you have satisfied yourself that you are well-prepared, presenting the material is a skill that will improve through practice.

17

Editing Your Writing

Editing itself is an excruciating act of self-discipline, mind-reading, and stable-cleaning. If it seems like a pleasure, something is probably wrong.

Arthur Plotnik, journalist and author (2005)

Key Concepts

- The editor's toolbox
- Structural editing
- Copy editing
- Final proofreading
- Common faults in English expression

Introduction

Whenever you write something, whether an essay, laboratory report, journal, or an article for publication, it begins as a rough draft. Through a process of reading and rewriting this rough draft is converted into a finished product. Editing is an essential part of the process of writing, and should not be seen as separate from it. Your first draft of any writing may be uneven, but once you have worked your way through to the final version it should be as good as you can make it within the limits of time, effort, and sanity. Oscar Wilde once said that he spent all morning trying to decide whether to put a comma in, and the whole afternoon deciding whether to take it out.

Each piece you write has a purpose and an intended audience, even if the only reader will be your tutor or lecturer. Never underestimate the work needed to bring an idea to fruition in good scientific writing. The essay on the life of gymnosperms may not be a novel, but that doesn't mean you don't need to give it your best effort.

Full editing is a three-stage process of structural editing, copy-editing, and proofreading. This chapter will briefly cover these three stages of editing as they apply to your writing, and will look at some common writing mistakes.

The editor's toolbox

The word 'standards' in editing has the specific meaning of a range of reference texts you can refer to for a particular document. This is particularly important for long documents, but you can apply it just as well to any writing assignment in your course. There are three types of standard reference texts worth consideration.

1   a recent, good-sized Australian dictionary and, if your first language is not English, a home-language–English dictionary is also recommended

2   an up-to-date subject-specific dictionary is an invaluable resource. You will have realised within the first week of your course that there is considerable terminology specific to each area of science that may differ to everyday language

3   a suitable text of writing style (such as *The Oxford Guide to Style*), which will include many useful tips for writing and editing.

Many online resources can also be very good and have the advantage of being free. Keep in mind, however, that you may need to trawl through vast amounts of information to find relevant information on the different styles, formats, and language use for your topics.

Structural editing

Once you have a version of your assignment that your think is the best one for the last draft, then it is time to begin the three-stage process of editing. The first stage is structural editing, also called *overview* editing. This is the big-picture view of your work. The key to understanding a piece of writing is to know its purpose and audience. If you take nothing else away with you from this chapter, remember to always question what the text is about and who it is for. With any piece of writing is it imperative to review the design of the piece. What is it trying to achieve? Can you name its purpose? All writing is (or should be) written with the readers in mind. At the undergraduate level perhaps the only reader is the person who will assess your work and award a grade for it.

One quick way to see whether you are on the right track with your piece of writing is to check your headings and subheadings. You can do this in *MS Word* in outline view. The Word program uses heading levels. You can change these styles to suit your requirements. Now you can tell whether you have the right order of information and you can check if there is enough content for each section. Are all your sections or paragraphs balanced? Have you overlooked some crucial information or dwelt too long upon another? One common fault is to repeat information, often in the slightly changed guise of another point, where really the material should come together.

Another problem sometimes occurs when you write material at different times, or, if in a report, different people have written separate sections. You need to make sure that the tone of the writing is consistent. In fact, aiming for consistency throughout the document is a prime aim of structural editing.

You need to ask:

* Who am I writing for?
* Is the format appropriate for its purpose?
* Are the aims defined?
* Is the shape tight (and word count correct)?
* Does it fit together?

You need to view the overall structure and:

* establish a hierarchy of titles
* check for balance
* ensure that the tone is consistent.

Tip

If you need to reduce your assignment's length, check each point or paragraph. Does each fit your purpose? Can you delete some and condense others? If you need to extend your assignment's length—check the outline. Are there subsidiary points you have not covered, but could? Have you apportioned the same amount for each point, or should some be expanded to match the others?

Copy-editing

The process of copy-editing is to check for clarity, logic, conciseness, and for impact. This is where you edit the:

* content
* comprehensibility of the arguments
* correctness, in that source material is from reliable sources
* consistency of argument and evidence
* clarity of the ideas being expressed
* compatibility of the conclusions with the evidence.

You must ensure the writing is balanced and that your sources are acknowledged accurately and completely (refer to Chapters 10 and 11). Your writing should acknowledge its antecedents, be valid, and add to the body of knowledge.

The comprehensibility of your writing is dependent on:

- an inviting beginning
- an even progression
- an authoritative conclusion.

The language is appropriate when it:

- is clear and concise
- is free of jargon (except specific terminology), unnecessary acronyms, and incomprehensible language
- maintains a constant tone and appropriate voice
- uses examples for support
- addresses its readers appropriately.

The conviction of the writing is determined by its

- worth
- ability to enlighten or educate
- ability to engage with current debate.

Remember, your own 'voice' sounds the best. Once you have written and submitted several papers you will begin to recognise that voice which is special to you. We all need to respect and honour our own writing voice.

Proofreading

Generally, we read our own writing as student writers rather than editors. We are not trained to proofread. In reading something for meaning we often skip over any errors. So for proofreading it is as if we need to put on a pair of different eyes, or a pair of special glasses, with which to read effectively. Read the sentence below and count the number of 'f's:

Frozen foods are the result of years of scientific study and the development of refrigeration.

How many did you get? It's easy to skip the terminal letters. There are seven occurrences of the letter 'f'. Often between the thoughts in your own head, the initial words on the page, and the final edited work, lots of hurdles exist. The journalist and editor, Plotnik, said,

An ingrained ear for language comes from reading good literature and balances the domestic babble, street talk, advertising drivel, and work jargon ravaging our brains.

To properly proofread you need to clear your head from all distractions. With your own work, time is of the essence. Don't try to proofread when you are multitasking (watching soccer, cooking dinner, downloading songs for your iPod); you need complete concentration. Imagine you are putting on a different hat, your proofreading hat. Preferably, work on an uncluttered space with an uninterrupted period of time to complete the entire job in one sitting.

TIP

Cover the page with a piece of card except for the lines you are working on. One way to do this is to cut a slot into a card or sheet of paper just large enough to fit the average sized sentence, then overlay this on your writing. This may be somewhat tedious and time consuming, but it is a simple way to 'trick' your eyes and mind to focus on a single sentence.

Proofreading involves checking each and every part of a text to ensure it is correct for:

- typographical errors (run the 'spellcheck' on your computer)
- spaces
- page length
- text that is aligned
- paragraphing that is consistent.

Common faults in English expression

Top spots for errors

Errors may crop up anywhere in your writing, but here are some good places to start looking.

- *Headings* are especially prone to mistakes, as are diagrams and tables.
- *Proper names* will cause much embarrassment if you get them wrong.
- *References* need to be complete, consistent, and accurate.
- *Cover page* needs the required and appropriate information.

TIP

Read your complete assignment out loud to make sure it 'sounds right' and that it makes sense. It is always a good idea to 'volunteer' someone else to read your work for you, but make sure it is someone who will give you an honest opinion.

Good English is not only correct, but it is also appropriate for the task. What is the correct sort of English for your work? If you are unsure about aspects of expression required for individual assignments, you will need to find out. You may have felt that good expression is only the preserve of arts subjects. Not so. All disciplines now care more about literacy involved in their own subject than before. Writing style is more than the sum of parts of grammar and punctuation. It includes tone, syntax, and word choice. It is incumbent upon you as a writer to submit work that is as correct as you can manage and as fully revised as possible.

Common error types

We are all prone to making mistakes in our writing, so one way to improve your writing is to be aware of the most common errors. Here is a list of the most common errors of expression in tertiary writing:

1   Subject–verb disagreement
2   Noun–pronoun disagreement
3   Ambiguous pronoun reference
4   Lack of parallel construction
5   Incorrect tense or change in tense
6   Sentence fragmentation
7   Misplaced modifier
8   Change in point of view
9   Wordiness
10 Lack of connectors
11 Colloquialisms, slang, or undefined abbreviations
12 Apostrophe misuse
13 Misuse of commas, colons, and semicolons
14 Incorrect heading formatting
15 Inconsistent paragraphing
16 Spelling errors
17 Misused words
18 Sexist or racist language
19 Strings of prepositions
20 Tautology

Some of these faults may sound familiar and no doubt all of us are guilty of some of these in our writing. If, on the other hand, some of these are unfamiliar to you but you wish to know more, then find the time to make yourself familiar with them. Once you understand the basic rules of writing your writing will improve.

When you write using a word processing program such as *MS Word*, the program can edit for you as you write. Spelling errors are underlined with red and grammar irregularities are underlined in green. If you are not sure what these error messages are highlighting, then investigate. Doing so may increase your English language skills and increase your confidence in using them.

Use of commas

The use and abuse of the comma takes up more space in books on writing than almost any other topic. Consider the following:

What's that ahead in the road?

What's that, a head in the road?

Commas can both separate and divide. If you are familiar with comma usage rules, you will be more confident as to whether you should or should not use commas.

The following is a list of examples of the many uses of commas. Find the example closest in structure to the one you need and use it.

- The original pub, a bluestone structure, is still standing.
- In all laboratory refrigerators except the Biology and the Chemistry ones, the fire hazards have been removed.
- No, the mid-year break has not yet arrived.
- Worried by the many complaints from employees, Worksafe began an investigation.
- The student's interjection, therefore, was ignored.
- I had heard the rumour before; consequently, I did not believe it.
- The laboratory coat, he insisted, was too tight.
- The middle bookshelf, for example, was undamaged.
- The crops—canola, wheat, and Salvation Jane—are in good condition.
- Three species of tree were observed, namely, eucalypt, acacia, and pine.
- The north side, which had been exposed to the sun, was badly faded.
- He moved to Tasmania, where the climate was not so humid.
- The people from Hawthorn, who had come early, obtained the best seats.
- To a trained forensic scientist, the problem would look easy.
- 'The tank is near empty,' she said, 'and the motorbike will stop soon.'
- The university is old, but it has been kept in good condition.
- Goats, alpacas, and ferrets are now selling for higher prices.
- We visited St Kilda, Lygon Street, and Phillip Island in Victoria; the Barossa, Fleurieu, and Clare regions in South Australia; and the Hunter Valley and Mudgee in New South Wales.
- The walls were painted red, blue, yellow, and black and white.
- The lecturer has a gentle, unassuming manner.

- The technician was a strong, young woman.
- Incrementally, imperceptibly, the creek wore away the bank.
- The Southern Vales, South Australia, is an important wine producing area.
- In 1995, 675 men were sent to jail.
- Please insert the name Kewell, Harry, in the proper place in the alphabetical list.
- June will be devoted to writing; July, to editing.
- Inside, the lecture theatre was more comfortable.

Avoiding tautology

A tautology is a repetition of the same idea, or using different words to say the same thing within the same sentence. A few simple examples will illustrate this:

- adequate enough
- meet together
- collaborate together
- mutual cooperation
- debate about
- new beginning
- follow after
- subsequently follow
- important essentials
- utterly unique.

Distinguishing meaning

It is important to use the correct term so that the word means what you think it means. A few examples could be:

- disinterested—impartial
- uninterested—not interested
- fewer—in number
- less—in amount.

Parallel structure

Make sure all the elements match, especially in lists. Read the introductory part of the sentence and make sure you match each element of the list with your introductory material. One of the best ways to do this is to ensure that each element of the list starts with say, a verb or a noun, but whichever you use make it consistent.

Summary

It may seem an obvious thing to say, but ensure that material to be edited is well organised. Keep all the files required carefully sorted. Before embarking on the editing process, read the whole document. Remember there are books that are essential to assist you in your writing or the assessment of your writing. At various stages of your study and later work, improvement with writing skills and the basics of editing will be of assistance.

This an introductory chapter only, and cannot provide all you need to know about editing, but you should now be able to self-edit to the best of your ability. You should now feel more confident to discern, judge, consider, estimate, reckon, regard, deem, appreciate, determine, decide, and conclude at least some things about writing you come into contact with. No piece of writing is too small or insignificant to benefit from your improved skills.

PART 5

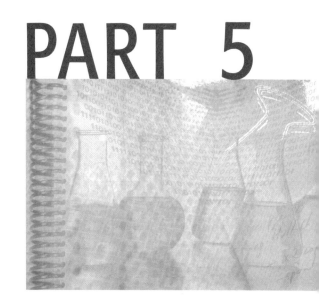

Quantitative
Methods

18

An Approach to Problem Solving

The formulation of the problem is far more often essential than its solution, which may be merely a matter of mathematical or experimental skill.

Albert Einstein, German theoretical physicist (1879–1965)

Key Concepts

- Blocks to problem solving
- An approach to problem solving
- A problem-solving flow chart
- Open and closed problems

Introduction

Problem solving is a skill we all have, and one that we continue to develop our whole lives. Your earliest problems in life may have been things such as how to get the attention of your parents to let them know you are hungry before you could speak, or how to walk without falling over. As you grew older you learned how to tackle more abstract problems such as the letters of the alphabet and how they can be used to make words, or learning how to count using numbers. Problems of all sorts continue to be a part of our daily lives.

Examples of everyday problems that all of us deal with

- *A game*: How can I beat my best friend at Dungeons and Dragons?
- *Getting something to work*: How do I set my DVD recorder to copy *Australian Idol*?
- *Making something happen*: How do I select the right numbers to win lotto?
- *Why something does not work*: Why won't my car start this morning?
- *Calculating something*: How much money do I need for food this week?

What do you do when you attempt to solve a problem? In general terms, when you solve a problem you use mental tools which allow you to examine the available data and devise a course of action.

Generally, science concerns finding solutions to problems. Scientific problems may range from finding a cure for dementia, saving the planet from the effects of pollution and global warming, to finding a better way to kill cockroaches. At the undergraduate level of your courses the type of problems you may be asked to solve will usually be a bit simpler, and could include such things as finding the pH of a buffer solution, calculating the age of a rock sample, writing a computer program to determine the rate of an enzymatic reaction, or determining the tensile strength of steel to build a bridge. This chapter looks at some of the strategies you could use to solve problems.

Blocks to problem solving

Generally, there are a number of reasons why you cannot solve a problem. If you have the intellectual skills, information, recognition of the desired outcome, and the necessary technical skills, you should be able to solve most problems. Therefore a failure to do so may lie elsewhere and could include any of the following:

* fear of failure to solve the problem
* a problem which is too difficult
* past bad experiences with this type of problem
* not enough time
* giving up too easily.

If you know something about the nature of the obstructions that impede you from solving a problem, you may be better able to deal with it. Past failures or the fear of failure may convince you that you cannot succeed, and therefore you may be afraid to even try. This then becomes a self-fulfilling prophecy. Is the problem really too difficult, or have you made some assumptions? You need to learn to look at the big picture but also be able to break things into small solvable problems. What parts can you do and which parts are causing the difficulty? Can you identify a particular problem step that is the sticking point? Why is this step more difficult, and what can you do to address the problem? Do you have all the information you need and the appropriate knowledge? If not, what can be done to get you to the point where you can tackle the problem? If the problem is a mathematics-based problem, do you lack sufficient skills? Lastly, have you spent sufficient time analysing the problem and how to solve it? It is easy to say it is too difficult without actually having spent sufficient time working on the problem. Sleep on it overnight and try again the next day when you are refreshed.

Much of science is practical or experimental, but all activity in science is essentially theoretical. That is, it requires you to think about a problem. As a physics lecturer once said, 'there is nothing so practical as a good theory.' A great deal of your learning will involve looking for solutions to problems. At the level of undergraduate science these problems usually have three distinct features:

1   The goal of the problem is usually clearly defined.

2   All the data required to solve the problem are provided or assumed knowledge.

3   The method of solution follows a recognised pattern.

Steps in problem solving

There are a number of strategies you can use when tackling any sort of problem. Like learning itself, problem solving is a process of trial and error. When attempting any problems, but particularly those that you find difficult or those that may not be familiar to you, you may need to develop a set of strategies by following a sequence, such as that shown below.

Clarifying the problem

1   Identify the nature of the problem.

2   Determine the value of the information or data provided.

3   Ask any questions that can help to clarify the required approach needed.

4   Define all necessary terms.

5   Identify any unstated assumptions or further information required.

TIP

Often the hardest part of any problem is to know how to get started. You may look at the question and decide that you do not know how to do this type of problem. This type of negativity is a self-fulfilling prophecy. Don't fall into this trap. Think of other problems you have solved that may be similar, no matter how remotely. Do other simpler problems first then come back to the one you cannot do. Try just playing around with the data with a pencil on a scrap of paper. If it does not make sense the first time around, leave it and come back to it later. Go for a walk. Sleep on it.

Devising a plan of attack

1   Draw a diagram that links the information or data given.

2   Decide whether you have seen a similar problem.

3   Think about lecture notes or other examples which may help.

4   Break the problem into a number of smaller problems.

5   Write down the steps for the solution of these smaller problems.

6   Generalise from these.

7   Look for relationships that connect these steps.

8   Develop a plan using these simple steps and test it.

TIP

A simple diagram of a problem may allow you to see the problem in a different way. Develop a diagram where you put what is known on one side, what you want to know on the opposite side, and any known relationships linking different variables and any other assumed information in the middle. Then think of ways to link the known to the unknown, using these known relationships.

Given information   What you want to find

A, B, C, D $\longrightarrow$ Y, Z

Known relationships

$$A \times B = C$$
$$N = M \div K$$
$$L = 6.023 \times 10^{23}$$

What else do you need that may be relevant?

K, M, N

Implementing the plan

1   Carry out the plan by completing each step.

2   Check your arithmetic at each step.

3   Look to see if the steps form part of the bigger picture.

4   Now match the diagram you devised earlier with the emerging picture.

Checking the answer

In checking your answer you will need to ask the following questions:

1   Is the solution reasonable?

2   Is the solution close to what you expected?

3   Can you verify the solution by an independent method?

4   Can you now apply your method to other problems of this type?

TIP

The use of modern calculators to do mathematically-based problems will often give an answer to many decimal places, but it cannot tell you whether the answer is *reasonable*. For example, you may have been asked to calculate the thickness of a cell wall. You have all the necessary data, and you are familiar with the required technique. You calculate the answer to be 3.125 cm. Is this a reasonable answer? The answer is clearly 'no', as this is not a reasonable answer. Cell walls are measured in micrometres, so your answer is much too large. Remember that the units of the answer are as important as the numerical value you calculate.

The answer to the question of how to solve a particular problem is often a matter of trial and error, the key aspect of which is to have a logical sequence with which to follow. Figure 18.1 shows one approach.

If the goal to solve a problem is clearly defined and the problem has a unique answer, it is called a *closed* problem. Generally, the typical problems you will be asked to solve at undergraduate level are examples of closed problems in which all the necessary data are given and the plan for solving the problem is likely to be familiar to the problem solver. Usually these problems involve the application of relationships, or manipulation of data using known methods.

There are many different types of problems which you may be asked to consider as part of your course. Most of these problems will fall into the category of *closed problems* for which there is normally a fixed answer. The sorts of problems you may be asked to solve could be of the type:

> *Find the pH of a buffer solution containing 0.01 M citric acid and 0.05 M sodium citrate. Citric acid has a pKa = 3.06.*

> *Taye–Sachs is a genetic disease. Two people who are both carriers of the disease, but do themselves not have it, have three children. What are the chances of at least one child having the disease?*

The problems listed below are not the sort of problems you may be asked to solve, but are exercises that explore your problem-solving skills and your ability to think logically. Some contain specific data, while others rely on your ability to make a 'guesstimate'.

Figure 18.1   A possible schematic approach to problem solving

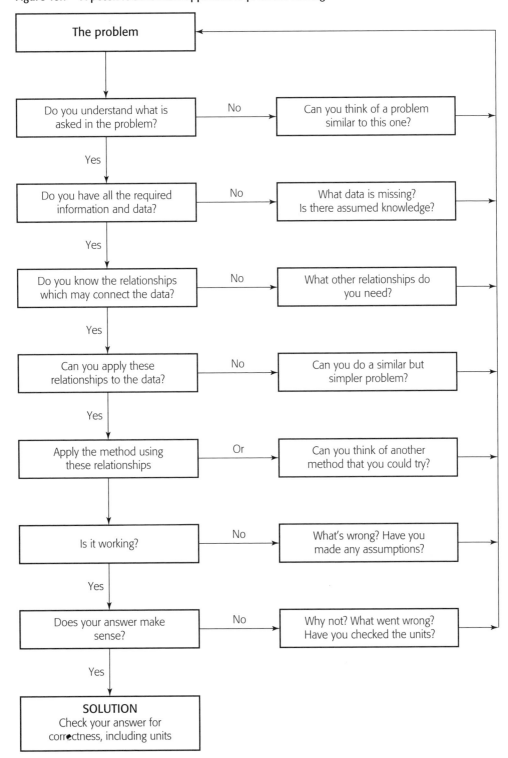

Examples of closed problems

1   A square milk crate can hold 36 cartons of milk. Can you arrange 14 cartons in the crate so that each row and column has an even number of cartons?

2   Arrange ten Christmas trees in five rows of four trees each.

3   How can a cylindrical block of cheese be cut such that there are eight equal portions using only three straight cuts?

4   What is the thickness of rubber left on the road by a car tyre for each revolution of the wheel? (This is more challenging as you have not been given any data.)

5   You have 27 metal balls, one of which is a different weight from the others. You also have a set of scales, which is a simple device which will allow you to determine only whether two objects are the same weight or whether one is heavier or lighter than the other. Using only four measurements, determine which ball is the one which is not the same weight as the others.

In comparison, open-ended problems or critical thinking tasks are designed to be different from the traditional closed problems with respect to the three features of undergraduate problems mentioned earlier. In most open-ended problems, part of the task has an open goal in the sense that it does not have a unique answer. Other examples may be creativity exercises that have wholly open goals. With a creativity exercise, you are given some information and are required to deduce or calculate as much as you can from the given data. You are encouraged to show all your deductions and calculations, no matter how unimportant they seem. Creativity exercises are designed to test your ability to identify goals and to use data to attain these goals. Some *open-ended* problems have missing data or even irrelevant data. You may be expected to identify and locate extra data or information you consider relevant in developing a plan to solve the problem.

Most *open-ended* problems relate to real-life experiences or issues. Because these problems are quite different in structure, content, and appearance from typical textbook undergraduate problems, the possible solutions are broad and variable.

Examples of open-ended problems

1   Can you solve the greenhouse effect?

2   How could you make your car run more cheaply?

3   Can you design the perfect golf ball?

4   Suggest a method of making more effective and safer pesticides.

Summary

The problems which form part of any undergraduate science course are often exercises by which you implement and practice the theories, techniques, and methods which are an integral part of each science discipline. Sometimes you will encounter problems where you are unsure at the outset what methods will be applicable, or precisely what the problem is. In many situations the uncertainty lies in the transition from the problem itself to a useful description of the problem. The more problems you solve and the greater the diversity of the nature of the problems, the better your problem-solving skills will become.

19

Fundamental Mathematics

Without mathematics the sciences cannot be understood, nor made clear, nor taught, nor learned.

Roger Bacon, English philosopher (1214–92)

Key Concepts

- Order of operations
- Scientific notation
- Exponents: powers and roots
- Logarithms
- Equations and dimensions
- The power of mathematics

Introduction

Imagine a life without numbers. There would be no way of knowing how much of a drug to give to a sick patient. There would be no way of describing how high the temperature was on a hot summer's day. You would have no need of a bank account as there would be no way of knowing how much money you have. There would be no speed limits and we would have to revert to telling the time using methods that did not need numbers.

Mathematics is a major foundation of our society but, like most foundations, it tends to be out of sight. This helps to create and perpetuate the notions that maths is hard, not much fun, and somehow mysterious. As a language, mathematics is grammatically very simple: it contains only a handful of nouns (the digits from 0 to 9) and even fewer verbs (add, subtract, multiply, and divide). The essential rules of its grammar are few and readily specified; all further rules arise as a logical consequence of combining these together with, perhaps, the introduction of the occasional convention so that we all agree on what is intended.

If you want to improve your proficiency with mathematics, you should try to *understand* what is being communicated rather than merely seeking to apply memorised rules. Whatever the mathematics you see or use, it should always make sense and mean something. It is reasonable to expect that most of the mathematics you encounter as a student should be 'do-able'. If it doesn't make sense then the missing element must lie in a misunderstanding of the mathematical language, or the way in which it is being applied. It is essential that you recognise when this has occurred and develop strategies to resolve the matter before proceeding. Ask yourself whether your problem lies in the equations or calculations you are trying to deal with, or whether the difficulty lies elsewhere in the subject matter.

This chapter is not an attempt to teach you mathematics, but rather to highlight some of the fundamental concepts of mathematics common to many sciences.

Order of operations

When performing calculations involving a number of operations, it is essential that you carry them out in the proper order. The correct sequence is referred to as BEDMAS, which is an acronym for:

Brackets Exponents Division Multiplication Addition Subtraction

That is, in evaluating a compound expression, terms in brackets must first be reduced to a single quantity (which, itself, may require the application of the BEDMAS sequence) or, alternatively, expand the expression to remove the brackets. Note that some expressions contain nested brackets (terms within brackets), in which case you must deal with the innermost brackets first and work outwards. Next, terms with exponents are processed (in effect, by carrying out the implicit repeated multiplication or division) and then any divisions are carried out before undertaking multiplication. Finally, you should arrive at an expression that is equivalent to the one you started with but now it will involve only terms to be added or subtracted. You should proceed from left to right until you arrive at the required answer—a number or quantity, or an expression in its simplest form.

While this description seems complicated, the BEDMAS procedure is a shorthand reminder. If you understand the language of mathematics you will realise that this sequence is the consequence of insisting that the evaluation of an expression makes sense, that any attempt to do things otherwise lacks meaning. Rather than relying on BEDMAS as a rule, you should aim to become fluent with mathematical language.

A more general version of this sequence is known as BODMAS, with the 'O' representing the broader concept of 'operations'. This includes not only exponents (powers) but also other mathematical functions such as roots, logarithms, and trigonometric functions (for example sine, cosine, tangent), all of which, like exponents, need to be resolved before evaluating the expression.

The BEDMAS/BODMAS sequence is automatically followed if you use a scientific calculator correctly, but not all calculators are created equal so it is very important that you take a little time to ensure that you know how to use *your* calculator.

Exponents—powers and roots

What do we do when we multiply two numbers? What does 'multiplication' mean? Consider a simple case, such as 3 × 4, which is shorthand for 3 + 3 + 3 + 3 or, equivalently, 4 + 4 + 4. That is, multiplication may be described as repeated addition.

This idea of repeatedly applying a fundamental operation is readily extended to the concepts of multiplication and division. This is what is involved in the introduction of *exponents*, also known as *powers*. Let us first consider repeated multiplication. When we talk about this, we need to specify which numbers are being multiplied. For instance, the quantity 7 × 7 × 7 × 7 is usually written as 7^4; the '7' is the *base* and the '4' is the *exponent* or *power*. This is a definition by example, an illustration that does not fully identify what is going on. In particular, it is not entirely clear what is being repeatedly multiplied, nor is it clear from the three multiplication symbols how the repetition is specified. Of course, you could say, 'It's obvious that four sevens are multiplied together'. But then, how would something like 7^0 be interpreted? Would you then say, 'This means that no sevens are multiplied together'? If so, what would this quantity represent? A better approach is to be more precise in the definition and write:

$$7^4 = 1 \times 7 \times 7 \times 7 \times 7$$

so that it now becomes clear that the number 1 is multiplied by the base number, as many times as specified by the number in the exponent.

Now it should be apparent that 7^0 is interpreted as 1 multiplied by 7, 0 times; that is, the 1 is not multiplied at all, and so $7^0 = 1$ (and by the same process any number raised to the power of 0 is 1). Similarly, $5^1 = 1 \times 5 = 5$, that is, any number raised to the power of 1 is the number itself. In practice, when expanding a number with an exponent into explicit repeated multiplications, the leading '1 ×' is dropped for convenience, since it doesn't contribute to the result.

More generally, and following the above example, in the expression b^c, the exponent is c and the number b is the base. Your calculator will probably have a button to evaluate powers of numbers. It may be marked $\boxed{x^y}$ or $\boxed{\wedge}$.

Multiplying and dividing powers

When quantities with exponents are multiplied or divided, the result can be simplified provided that both quantities have the same base. For example,

$$5^3 \times 5^2 = (1 \times 5 \times 5 \times 5) \times (1 \times 5 \times 5) = 1 \times 5 \times 5 \times 5 \times 5 \times 5 = 5^5 = 3125$$

$$6^4 \div 6^2 = (1 \times 6 \times 6 \times 6 \times 6) \div (1 \times 6 \times 6) = 36$$

However, if the quantities have different bases, no such simplification is possible. You can verify this by writing, say, $5^3 \times 6^2$ in expanded form as above; does the result look like 'repeated multiplication' in the sense intended by the definition of exponents? For essentially the same reason, while quantities in exponent form can be added and subtracted, an expression of that type cannot be simplified without evaluating the result unless both the base and the exponent are the same for each quantity. For instance, $3^2 + 3^3 = 36$, which cannot be expressed as a power of 3. One way of understanding this is that in the first case the two quantities are different: the first is a square (power of two) while the second is a cube (power of three), and you cannot add 'apples and oranges'. However, $3^2 + 3^2 = 2 \times 3^2$ is a valid simplification since the quantities being added are identical.

Powers of powers

Any number expressed as an exponential may itself be raised to a power. For example,

$$(5^3)^2 = 5^3 \times 5^3 = 5^{2 \times 3} = 5^6$$

You may, of course, verify this result using your calculator.

Negative powers

Exponents need not be positive numbers. They can also involve negative numbers. If we follow the original definition, given above, then how might 2^{-3} be interpreted? We know that positive powers represent repeated multiplication but you can't repeatedly 'un-multiply', that is, carry out a negative number of multiplications. The appearance of the minus sign in the exponent suggests that, instead, we need to carry out the opposite of multiplication, which is division. Thus:

$$2^{-3} = 1 \div 2 \div 2 \div 2 = \frac{1}{2} \div 2 \div 2 = \frac{1}{4} \div 2 = \frac{1}{8}$$

But $8 = 2^3$, so $2^{-3} = \frac{1}{(2^3)}$. That is, a negative power is the reciprocal or inverse of the corresponding positive power.

Roots and fractional powers

So far, we have looked at exponents that are positive or negative integers (whole numbers). Whole numbers are a special case of general number principles, and there is no reason not to broaden our view of exponents to include all numbers. The question then becomes one of how to interpret and use such quantities.

Consider 2^2 and 2^3: these are alternative ways to represent the numbers 4 and 8. Similarly, 10^2 and 10^3 represent 100 and 1000. Does it not seem reasonable that the number 5, say, could be represented in the form of 2 raised to some power, which logically must lie somewhere between 2 and 3? Similarly for, say, 208 can it be expressed as a power of 10, with the exponent lying between 2 and 3? Indeed, it is possible, but it is not easy to determine the precise value of the exponent (this involves logarithms, discussed later in this chapter).

If we restrict for the moment the range of fractions to be used in the exponent, the matter becomes clearer. But first, consider the following.

When 5 is squared we obtain 25. That is, $5^2 = 25$. The reverse of this process is called finding a *square root*. The square root of 25 is 5. This is written as $\sqrt{25}$. Note also that when -5 is squared we obtain 25, that is $(-5)^2 = 25$. This means that 25 has another square root, -5. In general, a square root of a number is that number which, when squared, gives the original number. There are always two square roots of any positive number, one positive and one negative, but *negative numbers do not possess a real square root*. Most calculators have a square root button, probably marked $\boxed{\sqrt{\ }}$.

The *cube* root of a number is the number which, when cubed, gives the original number. For example, because $4^3 = 64$ we know that the cube root of 64 is 4, written $\sqrt[3]{64} = 4$. All numbers, both positive and negative, possess a single cube root. Higher roots are defined in a similar way: because $2^5 = 32$, the fifth root of 32 is 2, written $\sqrt[5]{32}$. In general, the nth root of a number x is written $\sqrt[n]{x} = y$, where n is any number. Scientific calculators usually have a key to find the nth root of a number (it may be labelled $\boxed{x\sqrt{y}}$). No n^{th} root can be found for negative numbers unless n is an odd number.

Now, it turns out that $\sqrt{9} = 3$ can be written equivalently as $9^{1/2} = 3$, which is intuitively believable, since $9^0 = 1$ and $9^1 = 9$ so $9^{1/2}$ must lie between 1 and 9. Likewise, $\sqrt[3]{64} = 4$ can be written as $64^{1/3} = 4$; $\sqrt[5]{32} = 32^{1/5} = 2$, and so on. In general, the nth root of a number x can be expressed as $x^{1/n} = y$. There is no reason why fractional powers cannot be written as decimal fractions, so that, for example, $9^{1/2} = 9^{0.5} = 3$ and $32^{1/5} = 32^{0.2} = 2$. When roots are expressed in this form they can be readily evaluated on a calculator by using the same button that you use to evaluate powers, rendering the nth root key redundant.

This part illustrates that although we may give different names to the concepts of powers and roots, they are different aspects of the same underlying notion. These ideas are readily found to apply generally and are summarised in the rules of exponents (figure 19.1).

Figure 19.1   Rules of exponents

$$a^m \times a^n = a^m a^n = a^{m+n} \qquad\qquad a^{-m} = \frac{1}{a^m} \text{ and } \frac{1}{a^{-m}} = a^m$$

$$\frac{a^m}{a^n} = a^{m-n} \qquad\qquad a^1 = a$$

$$\left(a^m\right)^n = a^{m \times n} = a^{mn} \qquad\qquad a^0 = 1$$

In each of the above, a is the base and m and n are the exponents or powers. These can represent any values (subject to the constraints on roots of negative numbers), but it is essential to recognise that the base must be the same when the first two rules are applied.

Scientific notation

Many of the numbers used in science are very large or very small. For examples of large numbers, there are about 300 million sperm in a normal human ejaculate, the Milky Way galaxy contains approximately 400×10^9 stars and one second is defined to be the duration of 9 192 631 770 oscillations of a caesium-133 atom. For examples of small numbers consider that the radius of a hydrogen atom is about 6 million-millionths of a metre, the weight of a proton is about 1.5 million-million-million-millionths of a gram, and (in principle) physicists can measure or calculate processes that last as little as one trillion-trillion-trillion-trillionth of one second at a scale down to around one trillion-trillion-trillionth of a metre. If you lived in ancient Greece or Rome you would have great difficulty in dealing with these very large or very small numbers using Roman numerals in use at that time. Consider, for instance, a large number such as the distance to the moon from Earth, in miles. This distance, which is 238 857 miles, would have been expressed as:

CCXXXMMMMMMMMMDCCCLVII

Fortunately for them, perhaps, the Romans didn't know this distance, so writing it down was not a problem. A much more convenient way of expressing such a number is to use scientific notation, involving powers of 10, since this is the base of our counting system. For example:

$10 = 1 \times 10$	can be expressed as	10^1
$100 = 1 \times 10 \times 10$		10^2
$1000 = 1 \times 10 \times 10 \times 10$		10^3
$1\,000\,000 = 1 \times 10 \times 10 \times 10 \times 10 \times 10 \times 10$		10^6

Hence the distance to the moon would be expressed as 2.38857×10^5 miles. That is, the original number has been written as a multiple (2.38857) of the largest power of the base less than our number. In this case the power is 5, since $10^5 = 100\,000$ is less than 238 857. The next higher power is $10^6 = 1\,000\,000$ which is larger than our number. This convention uses a multiplier (in this case 2.38857) which is always a number larger than 1 but less than 10. Scientific notation is an alternative way of expressing a given quantity. Similarly, a small number can be expressed in terms of negative powers of 10, so that:

0.1	$= \frac{1}{10}$	can be expressed as	10^{-1}
0.01	$= \frac{1}{100}$		
	$= \frac{1}{(10 \times 10)}$	can be expressed as	10^{-2}

$0.001 \quad = \dfrac{1}{1000}$

$\qquad = \dfrac{1}{(10 \times 10 \times 10)} \qquad$ can be expressed as $\quad 10^{-3}$

$0.000001 = \dfrac{1}{1,000,000}$

$\qquad = \dfrac{1}{(10 \times 10 \times 10 \times 10 \times 10 \times 10)} \qquad$ can be expressed as $\quad 10^{-6}$

Hence we can represent a very small number, such as 0.0000000187, by counting the number of multiples of 10 from the place where we put the decimal point, which would then give us:

1.87×10^{-8}, or 18.7×10^{-9}, or 187×10^{-10}

Notice, again, that each of these is regarded as an alternative way of expressing the same number, 0.0000000187. While each is numerically equivalent, only the first case above is regarded as scientific notation because the multipliers in the last two are greater than 10. However, the second case is an example of a variation of scientific notation called engineering notation, where the power is chosen to be a convenient multiple of three so that the quantity represented corresponds to magnitudes of a thousand (or more) or one-thousandth (or less) in the metric system known as SI units, from the French term *Système International d'Unités*. Many of the commonly used exponents in the SI system are referred to by using a prefix or a symbol as shown in appendix 5.

So, say we have a plant cell membrane which is 0.0000032 or 3.2×10^{-6} m thick. We can also refer to this membrane as being 3.2 micrometres or 3.2 μm thick.

Addition and subtraction

When large or small numbers are added or subtracted, this can also be done by the use of exponents. Consider the following example.

$1.2 \times 10^{3} + 1.5 \times 10^{5}$

We have two numbers added together: 1.2 lots of a thousand added to 1.5 lots of a hundred thousand. To add these numbers, we need to change the way we describe them so that they are both expressed in the same terms. Now, 1.5×10^{5} is the same as 150×10^{3}, that is, 1.5 lots of a hundred thousand is the same as 150 lots of a thousand, hence $1.2 \times 10^{3} + 1.5 \times 10^{5}$ can be expressed equivalently as

$1.2 \times 10^{3} + 150 \times 10^{3} = 151.2 \times 10^{3}$

$= 1.512 \times 10^{5}$

Note that a common mistake with addition like this is to add the exponents. It should be straightforward to convince yourself that this doesn't work. Consider the example:

$$10^3 + 10^5 = 1000 + 10\,000 = 11\,000 \neq 10^8$$

Multiplication and division

Just as exponents can be used to add and subtract numbers, they can be used to multiply and divide, provided that the base numbers are the same, as they will be when dealing with scientific notation, since the base will always be 10. The exponents are added for multiplication and subtracted for division, following the rules for exponents. Consider the following:

$$(1.27 \times 10^3) \times (2.89 \times 10^5)$$

This can be written as

$$1.27 \times 2.89 \times 10^3 \times 10^5$$

since the order of multiplication does not change the result (for example $2 \times 3 = 3 \times 2$). Now, recalling the rules of exponents,

$$10^3 \times 10^5 = 1000 \times 10\,000 = 10\,000\,000 = 10^{3+5} = 10^8$$

so

$$1.27 \times 2.89 \times 10^3 \times 10^5 = (1.27 \times 2.89) \times 10^{3+5}$$

$$= 3.67 \times 10^8$$

Similarly:

$$\frac{3.29 \times 10^6}{6.08 \times 10^9}$$

$$= \frac{3.29}{6.08} \times 10^{6-9}$$

$$= 0.54 \times 10^{-3}$$

$$= 5.4 \times 10^{-4}$$

Logarithms

Logarithms are another useful and powerful way of expressing large and small numbers. An example of this is the use of logarithms to express the concentration of hydrogen ions in chemistry, which is called the pH scale. Other examples where logarithmic scales are used include chemical reaction rates, the rate of action of enzymes, and the rate of radioactive decay.

Besides their convenience as a way to express and manipulate large and small numbers, logarithms are frequently used in science to compress data, mimicking behaviour found in nature. For example, the human ear can accommodate a huge range of sound intensity so that we can cope with, say, the roar of a jet engine or a rock band yet still hear a pin drop or the sound of the merest whisper. A nearby jet engine produces four times the sound intensity of one that is twice the distance away but this difference doesn't equate to four times the perceived loudness. The same difference in loudness may be perceived when we double our distance from someone speaking in a normal voice, yet the change in sound intensity is vastly different from that encountered from a jet engine. Whereas sound intensity as a function of distance follows a power law (in this case the inverse square law), our perception of loudness follows an approximately linear scale. The pictures below illustrate this behaviour: inverse square behaviour is shown on the top plotted on a linear scale, whereas the graph on the bottom shows the same data plotted on a logarithmic scale, so that the curve of the former appears as a straight line (figure 19.2).

Figure 19.2   Plotting logarithms

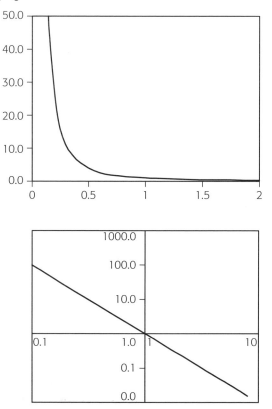

This is a common technique for analysing data from laboratory experiments—when data points appear to fall on a curve, either the data are re-plotted using logarithmic scales or the logarithms of the data are plotted on regular scales (sometimes a combination of both techniques is employed). The conversion to linear form can make analysis easier and may reveal relationships in the data that otherwise might not be obvious.

What are logarithms?

Logarithms are used to write, in a different form, expressions involving powers. In the same sense that multiplication and division are opposite, logarithms are the opposite of powers (powers, particularly to base 10 and base e, are sometimes called *antilogarithms*). If you can work confidently with powers, you should have no problem handling logarithms.

Consider the equation, $100 = 10^2$, which describes a true relationship between 100, 10 and 2. We can express the relationship equivalently as:

$$\log_{10} 100 = 2$$

This is read as 'the logarithm, to base 10, of 100 is 2'. As another example, since $2^5 = 32$, we could write $\log_2 32 = 5$ ('the logarithm, to base 2, of 32 is 5'). More generally,

$$\text{if } a = b^c, \text{ then } \log_b(a) = c$$

In principle, the base may be any positive real number (other than 1) but, in practice, logarithms are calculated using only a few common bases. Most frequently, bases are 10 and e, which represents the *exponential constant*, with a value of 2.718 (to three decimal places). It is used because it is a number that occurs in the mathematical description of many natural phenomena, particularly those that involve growth or decay, such as population change, heating and cooling, radioactivity, and compound interest. Logarithms to the base e are referred to as natural logarithms, and expressed as $\ln x$.

Using logarithms, it becomes straightforward to answer a question mentioned previously. We have seen that it is possible to represent a number such as 208 as a power of 10, and we know that the exponent should lie somewhere between 2 and 3 (since $10^2 < 208 < 10^3$). But what is the value of the exponent? Using logarithms,

$$10^x = 208, \text{ so that } x = \log_{10}(208)$$

A scientific calculator will calculate logarithms to bases 10 and e, using buttons labelled 'log' for base 10 and 'ln' for base e. In practice, whenever you see $\ln(x)$, it refers to the natural logarithm; that is, $\ln(x)$ means the same as $\log_e(x)$. It is common usage that whenever you see $\log(x)$, without the base explicitly shown as a subscript, this means the base is 10. That is, $\log(x)$ means $\log_{10}(x)$.

It is also common to encounter logarithms written without brackets, although this may lead to misunderstanding. For example, log x is clearly intended to mean $\log(x)$, but what of log $x.y$? Does this mean $\log(x.y)$ or $\log(x).y$? While it may be possible to infer the correct meaning from the context, this is often not the case. The safest way to make the meaning clear is always to use brackets.

Laws of logarithms

The laws of logarithms allow expressions involving logarithms to be written in various ways. The laws apply to logarithms of any base but note that *the same base must be used throughout a calculation.*

First law of logarithms

Consider the equation $c = a \times b$. We can write each quantity as a power of 10. That is, we can write:

$$a = 10^x, b = 10^y, c = 10^z$$

so that

$$10^x \times 10^y = 10^z, \text{ where } z = x + y \text{ (from the rules of exponents)}.$$

But we also know that

$$x = \log(a), y = \log(b), z = \log(c)$$

so $z = x + y$ is re-written as

$$\log(c) = \log(a) + \log(b)$$

However, we started with $c = a \times b$ and so we find that

$$\log(a.b) = \log(a) + \log(b).$$

This is known as the first law of logarithms.

Second law of logarithms

Similarly, we could start with $c = \frac{a}{b}$ and proceed to find

$$\log\left(\frac{a}{b}\right) = \log(a) - \log(b).$$

You should verify this for yourself.

Third law of logarithms

By setting $b = a$ in the derivation of the first law, we find that

$$\log(a.a) = \log(a^2) = \log(a) + \log(a) = 2\log(a)$$

Similarly, setting $b = a^2$, we would find that $\log(a^3) = 3\log(a)$ and thus, in general:

$$\log(a^n) = n\log(a)$$

Finally, note that in any base $\log(1) = 0$ and also that $\log(0)$ is undefined, as is the logarithm of any negative number.

Equations and dimensions

There are many instances in science where the interrelationship between variables is expressed as an algebraic equation. For example, in chemistry the relationship between certain properties of gases is summarised by the ideal gas equation:

$$PV = nRT$$

where P is pressure

       V is volume

       n is amount of gas

       R is the ideal gas constant, and

       T is the temperature (in K)

You may be required to be rearrange an equation to change the subject. For example, if the pressure were the desired subject of the above equation, this can be achieved by dividing both sides of the equation by the volume, V. Thus

$$\frac{PV}{V} = \frac{nRT}{V}$$

and $P = \frac{nRT}{V}$

If you want T to be the subject of the equation, you need to remove nR from both sides of the equation by dividing both sides:

$$\frac{PV}{nR} = \frac{nRT}{nR}$$

hence $T = \frac{PV}{nR}$

Units and dimensions (dimensional analysis)

An essential aspect of equations in science is that they describe relationships between the quantities involved and, as such, must always make sense. It should be possible to translate the mathematical language into equivalent (though perhaps not so precise) English language. One key to ensuring the sense of equations is to recognise the literal meaning of the '=' sign: the quantity on the right-hand side (RHS) must be precisely equivalent to the quantity on the left-hand side (LHS). The validity of an equation can be readily checked by carrying out *dimensional analysis* and for this there are two simple rules—quantities can be added or subtracted only if they have the same units (dimensions), since it does not make sense to add, for example, kilograms to metres; and the quantities on each side of the equals sign must have the same dimensions. To carry out a dimensional analysis of the previous equation, for example, we replace each variable with its dimensional units (written in square brackets):

T is in Kelvin [K], P in kilopascals [kPa], V in litres [L], n in moles [mol], and R in the derived units [kPa L^{-1} mol^{-1} K^{-1}]. Thus

$$T = \frac{PV}{nR} \rightarrow [K] = \frac{[kPa][L]}{[mol][kPa.L.mol^{-1}.K^{-1}]} = \frac{[\cancel{kPa}][\cancel{L}]}{[\cancel{mol}][\cancel{mol^{-1}}][\cancel{kPa}][\cancel{L}][K^{-1}]} = [K]$$

Dimensional analysis can also be helpful if you want to describe derived units in terms of basic quantities. For example, there are seven SI quantities used to describe the physical world. These are listed in appendix 5.

Pressure means the amount of force per unit area, so one pascal is defined as one newton per square metre (1 Pa = 1 Nm^{-2}). Force is the product of mass and acceleration. Now, kg, m, and s are all base units and so pressure has units of:

$$Pa = N.m^{-2} = kg.m.s^{-2}.m^{-2} = kg.m^{-1}s^{-2}.$$

Dimensional analysis can tell you that an equation is wrong but it can't be used to prove that an equation is correct, since it cannot account for dimensionless constants (such as factors of ½ or π). However, it will let you quickly check whether you have used the correct units for something like the ideal gas constant, R, or the form of an equation that you are unsure about.

Let us look at an example from medicine. In drug calculations, it is necessary to calculate the volume (V) in millilitres (mL) of a solution to deliver the correct dose (D) in milligrams (mg), and the solution supplied as stock (S) in milligrams per millilitre (mg/mL). To calculate the volume of the dose, if you can't remember whether the equation should be V = S/D or V = D/S, then since the volume required (the LHS of the equation) will be in millilitres, the RHS must also be in millilitres. A dimensional check of the first formula gives:

$$vol = \frac{stock}{dose} \rightarrow [mL] \stackrel{?}{=} \frac{[mg/mL]}{[mg]} = \frac{[\cancel{mg}]}{[\cancel{mg}] \times [mL]} = \frac{1}{[mL]}$$

Clearly, the RHS and LHS are not the same, whereas

$$\text{vol} = \frac{\text{dose}}{\text{stock}} \longrightarrow [mL] = \frac{[mg]}{[mg/ml]} = \frac{[\cancel{mg}]}{[\cancel{mg}]} \times [mL] = [mL]$$

confirms that this is the correct formula.

The ability to work confidently with equations and formulas, both algebraically and numerically, is an essential tool for all science students, regardless of which branch of science pursued.

For a final illustration of the usefulness and power of mathematics, consider the following example.

Example: The A bomb

The first nuclear weapon used in a war was dropped by the Americans on the Japanese city of Hiroshima on 6 August 1945. The bomb, called *Little Boy,* had the following statistics:

Weight	8970 lb (4078 kg)
Length	10 ft (3 m); diameter: 28 in (70 cm)
Fuel	Highly enriched Uranium-235; 'Oralloy'
Uranium fuel	Approximately 140 lbs (63 kg)
	Uranium target component ferried to the island of Tinian via C-54 aircraft of the 509th Composite Group
Delivery:	B-29 Enola Gay piloted by Col. Paul Tibbets
Efficiency of weapon:	Poor (approximately 1.38%)
Explosive force:	20 000 tons of TNT (estimated maximum equivalent)

The energy produced by the blast was determined as equivalent to that produced by 12.5 kilotons (12 500 tons) of TNT. Now, 1 ton of TNT yields 4.184 x 10^9 J and so 12.5 kilotons will yield energy amounting to:

$$E = 12.5 \times 1000 \times (4.184 \times 10^9 \text{ J}) = 52.3 \times 10^{12} \text{ J}$$

Alternatively, in base units, $E = 52.3$ x 10^{12} kg m²s⁻² since joules are derived dimensions. Einstein's famous equation $E = mc^2$ can be rewritten as

$$m = \frac{E}{c^2}$$

The speed of light, c, is 2.9979250 x 10^8 msec⁻¹, so $c^2 = 8.99755$ x 10^{16} m²sec⁻² and, using the value for E from above, we can calculate the mass of uranium that was converted to energy in the blast:

$$m = \frac{52.3 \times 10^{12} \text{Kg.m}^2\text{s}^{-2}}{8.99755 \times 10^{16}\text{m}^2\text{s}^{-2}} = 5.8 \times 10^{-4} \text{ kg or } 0.58 \text{ g}$$

This is about the mass of two aspirin tablets.

Now for this atomic bomb, most of the 63 kg of uranium-235 was converted to lighter elements (krypton-92 and barium-141) and only 0.58 g was converted to energy. So you can see that if a small nuclear reaction could be contained and the amount of energy released slowly, it could be used to light a city for weeks, power a nuclear submarine for years or a fleet of cars for decades.

Summary

This chapter has been a brief introduction to some of the mathematical issues that arise in all sciences. Even the most descriptive areas of study use numbers to describe patterns and relationships. So if your mathematical skills are not adequate, your progress may be impeded. If you are unsure of the required level of maths skills, seek advice from your course coordinator about the mathematical skills expected of students.

20

An Introduction to Calculus

Mathematicians are like Frenchmen: whatever you say to them, they translate it into their own language, and forthwith it means something entirely different.

Johann Wolfgang von Goethe, German polymath (1749–1832)

Key Concepts

- What is calculus?
- Differential calculus

- Integral calculus

Introduction

Many of the concepts in science require an understanding of the rate of change of an event. For example, if you were a chemist you might be interested in developing new drugs that could impact on the rate of reproduction of cancer cells. Similarly, as a biologist working with enzymes, you may have observed a particular rate of action of an enzyme on the addition of a particular catalyst, and now want to determine what the magnitude of this change is if the rate could be monitored for a particular period. The same principles could apply if you were an oceanographer observing the changes in ocean current temperatures, or a medical scientist recording the impact of a new kidney dialysis technique. What each of these hypothetical examples has in common is that the changes observed or predicted can be deduced by mathematics, in particular the use of calculus.

Calculus (strictly, 'the calculus') has its foundations in antiquity, dating at least from the time when Archimedes devised extraordinarily clever techniques to calculate the area enclosed by a curve. Modern calculus is attributed jointly to Isaac Newton and Gottfried Leibniz who were bitter rivals in mathematics, and who independently showed how to calculate the slopes of curves and find instantaneous rates of change. Like so many other areas of mathematics, calculus has two major aspects, each in some sense 'opposite' to the other. One is called *differential calculus* and the other *integral calculus*.

Differential calculus

Differential calculus is used to determine the instantaneous rate of change of algebraic and trigonometric functions and also their minimum and maximum values, or turning points. For example, we might write the equation $y = x^2$ to describe a parabola (the shape of the reflector in a torch is an example of a parabolic curve). If we plot this curve, we observe that the y-value decreases as the x-value increases, but at a reducing rate until both x and y are zero. Then the curve changes direction and the y-values increase with increasing x-values, and at an accelerating rate, as shown in figure 20.1.

Figure 20.1   Parabolic curve

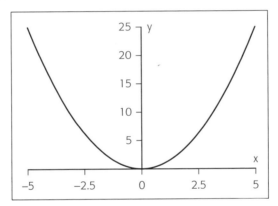

In such a graph, we have a sense of the rate of change of y with respect to x when we think of how steep the curve is at various places. If this were a valley, we would find it easier to walk around near the bottom of the curve than try to climb the sides. The idea of steepness follows the notion of the gradient or slope of a ramp, often described as the ratio 'rise over run' (figure 20.2).

Figure 20.2   Gradient of a straight line

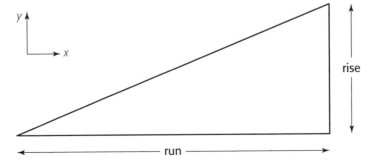

Now, if rise is defined as the y-direction (vertical) and 'run' as the x-direction (horizontal), then 'rise over run' can be written as:

$$\text{slope} = \frac{\text{rise}}{\text{run}} = \frac{\Delta y}{\Delta x}$$

where Δy is read as delta-y, meaning change in y-value, and Δx represents a change in x-value. The slope or gradient of any straight line can be quantified in this manner, but what of a curve such as $y = x^2$? We can draw a straight line between any two points (A and B) on the curve, as shown below, and find its slope, but what does that tell us about the steepness of the curve between those points (figure 20.3)?

Figure 20.3   Gradient of a curve

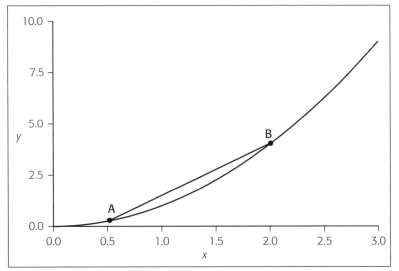

Close to A, the curve is less steep than the line, whereas close to B, it is steeper. The best we can say is that the steepness of this line tells us the average gradient of the curve between A and B. You can think of this in the same way as you understand average speed when travelling from one place to another by car. However, when driving a car, calculating the average speed can be useful for planning a journey, but once on the road what we need to know is our *actual* or *instantaneous* speed at any given time, in order to stay within the designated speed limit.

Speed means a rate of change of position; that is, how position changes with respect to time. Average speed is the change in position over some interval of time. Actual speed means the rate of change of position at any instant in time. Referring back to the graph, this corresponds to determining the steepness of the curve at any point rather than over some interval A to B. This can be found by drawing a tangent to the curve at the point of interest, where a tangent is a straight line that touches the curve at one point, and then finding its gradient or steepness. This works for us because it can be shown that for any given point on a curve, there is only one tangent. Figure 20.4 illustrates this idea, showing the tangent to the curve at point A.

While we can, of course, measure the slope of the tangent drawn at various points, this may not be convenient. In particular, what Leibniz and Newton developed was a means of obtaining a *formula* for the gradient when the formula that specifies a curve is known.

Broadly speaking, the answer lies in constructing the line between the point of interest, A, and another arbitrary point, just as we did to obtain the average slope. Then, move the second point down the curve towards the first, effectively making both Δx and Δy smaller, as small as you like. As the points move closer together, the slope of this shrinking line gets ever closer to the slope of a tangent at A. Make Δx *infinitesimally* small and the gradient coincides exactly with that of the tangent.

Figure 20.4   Tangent to a curve

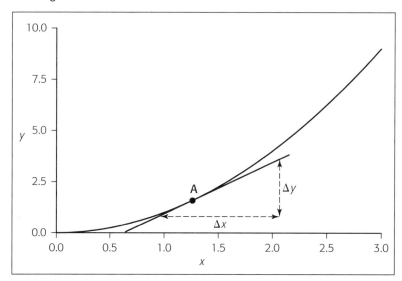

Mathematicians have adopted a notation to describe what happens. They say that

$$\frac{\Delta y}{\Delta x} \rightarrow \frac{dy}{dx}, \text{ meaning 'in the limit as } dx \text{ goes to zero'.}$$

This is read as delta-*y* over delta-*x* tends to *dy* over *dx* in the limit. What has happened is that the Greek letter delta Δ, which is used to denote a *macroscopic* difference, is replaced by the letter '*d*' to denote an *infinitesimal* difference.

TIP

This notation is frequently a source of confusion for students who haven't studied calculus. In particular, *dy/dx* should be read as a single identity; *d* is *not* a variable or pronumeral and doesn't represent a number to be multiplied.

Using this notation, if we are given the equation $y = x^2$, say, then we write

$$\frac{dy}{dx} = \frac{d}{dx} x^2 = 2x$$

to denote the instantaneous rate of change of y with respect to x. The quantity $2x$ is called the *derivative* of x^2. At $x = -2$ the slope of $y = x^2$ has the value $2 \times (-2) = -4$ (a line with negative slope goes 'downhill' when viewed from left to right). At the point $x = 0$, the slope is also zero (this is the bottom of the curve, the turning point). Likewise, at the point $x = 3$, the gradient is $2 \times (3) = 6$ (indicating that the curve is sloping upwards in the positive direction).

In general, the quantity dy/dx is referred to as the derivative of y with respect to x. Similarly, if a quantity V (volume) depends on another quantity t (time), then the quantity dV/dt is called the derivative of V with respect to t and represents the instantaneous rate of change in V with t.

Of course, this is useful when we know how to calculate the derivative. This process is called *differentiation* (hence 'differential calculus') and there are many standard techniques for differentiating algebraic and trigonometric functions.

Differentiation may be applied repeatedly. We can take the derivative of a derivative, known as the *second derivative*, and this is denoted as

$$\frac{d^2y}{dx^2}$$

(read as 'd-2-y by d-x-squared'). If the original equation describes a change of position with time, the first derivative is speed, and the second derivative gives the acceleration (which is the rate of change of speed). That is, the first derivative is to the original function as speed is to distance; the second derivative is to the first derivative as acceleration is to speed.

A variation of this notation is used when equations are expressed as functions. For instance, instead of writing $y = x^2$, say, which emphasises the relationship between x and y, we could write $f(x) = x^2$, where the 'f' is a label for a function that, given a number as input, produces the square of that number as the output. Then the derivative of that function may be written equivalently as

$$\frac{df}{dx}, \text{ or } \frac{d}{dx} f(x)$$

or more compactly as $f'(x)$ (read 'f-dash x'). In this notation, a second derivative would appear as $f''(x)$ (read 'f-double-dash x'), and any higher derivative is denoted by an extra dash.

The derivative as described above is used for functions or equations with only one independent variable. Where there are more than two variables, the notation is altered to identify that the *partial derivative* is determined. For instance, in the ideal gas law,

$$PV = nRT \Rightarrow P = \frac{nRT}{V}$$

(the $\Rightarrow$ means 'implies that') the pressure P of a given quantity n of an ideal gas depends on the temperature T and volume V of the gas. The relationship could be rewritten as a function, $P(V,T) = \dfrac{nRT}{V}$, to make this dependency clear. Then

$$\frac{\partial P}{\partial V} \text{ and } \frac{\partial P}{\partial T}$$

denote the partial derivatives of P with respect to V (when the temperature is held constant) and with respect to T (when the volume is held constant), respectively. These are read as, for example, 'd-P by d-V', and the symbol ∂ distinguishes the partial derivatives of functions of several variables from the derivatives of single variable functions.

Derivatives of many standard functions are often given in elementary undergraduate science texts and the various techniques of differentiation are standard content in first-year university mathematics courses.

Integral calculus

We have seen that equations and functions describe how quantities are related. We have also seen how the rate of change can be determined by differentiation—that is, by using differential calculus. On the other hand, there are many occasions when we might want to determine the magnitude of change. This is integral calculus. For example, suppose we observe an object moving at a uniform speed of 60 kmh⁻¹ or 16.67ms⁻¹, and plot speed as a function of time, giving the straight line graph on the left in figure 20.5.

Figure 20.5   Graph of speed vs time, zero acceleration

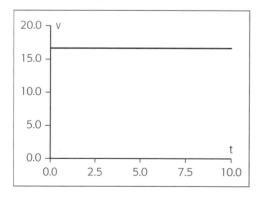
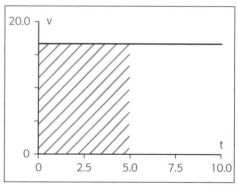

We could ask, 'How far has this object travelled after five seconds?' Since average speed is given by $v = \frac{s}{t}$, where s is the distance travelled and t is the elapsed time, it follows that the distance is the product of speed and time: that is, $s = vt$. This product corresponds to taking the height of the line in the graph (the value for v on the vertical axis) and multiplying it by the appropriate segment of the horizontal axis, which is the area of the shaded region of the right-hand part of figure 20.5. We find this to be

16.67 ms^{-1} × 5s = 83.33 m, which is the distance travelled in five seconds. That is, the distance travelled is equivalent to the area under some portion of the speed vs time graph.

Can this idea be applied in other situations? Consider the case of uniform acceleration, so that the speed vs time graph is still a straight line but now with non-zero slope. Figure 20.6 shows the situation with a constant acceleration of 2.5 ms^{-2}.

Figure 20.6   Graph of speed vs time, non-zero acceleration

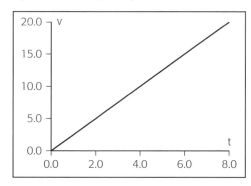
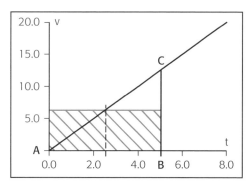

The area under the portion of the graph from $t = 0$ s to $t = 5$ s is the area of the triangle ABC on the graph on the right, which is 0.5 × 12.5 ms^{-1} × 5s = 31.25 m (the area of a triangle is half its height times the base). This corresponds to the area of the shaded rectangle, whose top edge intersects the graph halfway at $v = 6.25$, $t = 2.5$, that is, 6.25 ms^{-1} × 5s = 31.25 m.

Notice that in this example the average speed corresponds to 6.25 ms^{-1} and the object has moved at that speed for five seconds to cover a distance of 31.25 m. Again, the distance travelled is equivalent to the area under the relevant portion of the graph.

It is straightforward to show that this must always be the case and the problem becomes one of finding a method of determining areas in specific instances over specified intervals. The goal is to find a technique to cover the general case.

This is what *integral calculus* does. It provides methods and tools to find the *anti-derivative* of many functions.

Before introducing the notation relevant to this part of the language of mathematics, it will help to extend the previous concepts further. Look at the top curve in figure 20.7. At first glance, while it might be reasonably straightforward to estimate by the use of rectangles the area under the curve, say between $x = 2.5$ and $x = 3.75$ (where the slope is almost constant), it may not be immediately apparent how we might extend the method to find the area over any arbitrary interval. The trick is to cover the area with rectangles of width Δx whose heights correspond to the average y-value over each Δx interval. Then it is a matter of finding the area of each rectangle and adding all the areas together.

The result will be an approximation of the area under the curve.

Figure 20.7   Finding the area under a curve

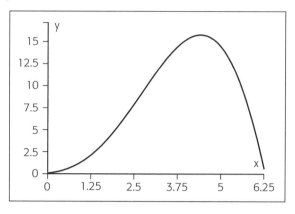

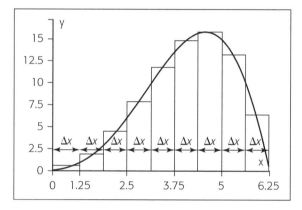

The result will be an approximation of the area under the curve but it can be made close to the actual value by taking the width of each rectangle to be smaller (of course, that means there will be many more rectangles). If we continue in this manner, so that the width of each rectangle becomes infinitesimal, then $\Delta x \longrightarrow dx$ (the 'dx' has the same interpretation as in differential calculus) and the height of each rectangle will be specified by the function that defines the curve at the location of the rectangle. When the area of these rectangles are added, the result will be the area under the curve.

The notation involves the equation or function (for the rectangle heights), the dx (or its equivalent, according to the name of the independent variable, for the rectangle widths), a symbol to indicate summation (so that all of the areas are added) and, optionally, start and end values to indicate in which portion of the curve our interest lies.

Specifically, if $f(x)$ is a function that specifies the curve, then

$$\int_a^b f(x)dx$$

specifies the area between $f(x)$ and the x-axis, between over the interval from $x = a$ to $x = b$. This is called the *definite integral* of $f(x)$ and a is the *lower limit*, while the b is the *upper limit*. Notice that the integration symbol $\int$ looks rather like the letter 'S' for 'sum', which is how it originated. If the limits are omitted, then

$$\int f(x)dx,$$

is called the *indefinite integral*, or *anti-derivative* of $f(x)$. The difference between the two forms is considerable. The definite integral always represents a number (the magnitude of the area under the curve), whereas the indefinite integral is a *function*.

To illustrate, recall that the derivative of $f(x) = x^2$ is $\dfrac{df}{dx} = 2x$; let us now take the '$2x$' and define $g(x) = 2x$. Next, we find the indefinite integral of $g(x)$:

$$\int g(x)dx = \int 2x\,dx = x^2 + c = f(x) + c$$

from which we obtain the result which is the original function plus an arbitrary constant, c.

As with differentiation, standard integrals are often published in undergraduate science texts and the techniques of integration can be found in any first-year calculus course.

Summary

Any scientific study which involves observations of the change of one variable with respect to another, or the measure of the magnitude of a change, can be assisted through the application of these techniques. You may never use these skills in your studies, but knowing just a little about them will help to broaden your outlook. The main point is that you should recognise calculus notation and have a basic understanding of what it means. The rest is detail that you can attend to and practice elsewhere. A table summarising some of the functions commonly used in calculus is shown in the appendices.

21

Basic Statistics

According to statistics, if you stand with one foot in a hot oven, and the other in a bucket of ice, you should feel perfectly comfortable.

Bobby Bragan, US sportsman and coach.

Key Concepts

- Why use statistics?
- Categorising data to identify patterns
- The use of graphs

- Lines of best fit in graphs
- Correlation between variables
- Using correlation coefficients
- Statistical terminology

Introduction

Here is an interesting statistic. Of all the road deaths in Australia each year, 30% of those who die have a blood alcohol reading of greater than 0.05 (the legal limit). This would be a good reason not to drink and drive. On the other hand, it also means that 70% of road deaths are people who have not been drinking. Does this mean that you reduce your statistical chances of dying in a road smash by drinking alcohol?

Now consider another statistic. During the Vietnam War, extensive data were collected on American casualties. The demographic data of soldiers killed in action (KIA) included, among other things, age, ethnic group, height, weight, and cause of death. One of the interesting statistical analyses of the data revealed was that there was a positive correlation (a linear association) between shoe size and death rate. The larger the shoe size, the higher the death rate. Can we conclude from this analysis that soldiers with large feet are more likely to die in battle?

So why do we use statistics? The answers to this question are many and varied. In the majority of science-based professions where practical, experimental, or survey work is involved, an important feature is the collection, collation, and analysis of data. Raw data may be collected in many different forms, and thus there is often a need to manipulate it so that it can be interpreted.

Statistics are used by scientists to evaluate the results of medical trials on the effectiveness of a new drug; by governments to monitor population trends to meet the demand for infrastructure, such as public transport, hospitals, and schools; to investigate the relationship between factors such as peer pressure and youth suicide, or between obesity and cardiovascular disease.

This chapter has a simple objective, which is to introduce some basic statistical terminology and methodologies used in many sciences, with the caveat that we can do little more than just mention some basic concepts. Statistics is an area of study where a basic understanding of mathematical principles is imperative. These days statistical analysis is simplified by the use of computer programs such as SPSS, MINITAB, or S-PLUS.

Categorising data

One of the simplest ways in which raw data can be made more intelligible is to categorise it into groups based on defined categories. As an example, consider that you are a marine biologist working for the Tasmanian Department of Sea Fishing and Aquaculture on a project collecting data on the number of blue-fin tuna caught in a given month, within a certain depth of ocean off the Tasmanian west coast. Each tuna caught is tagged with a label indicating the depth at which the net was set. The raw data could look something like the following:

Raw data

Tuna (in 100 kg lots) caught at the following depths in metres off the west coast of Tasmania for April 2005 (N = 38)

75, 67, 76, 71, 73, 86, 72, 77, 80, 75, 80, 96, 93, 75, 73, 83, 81, 82, 73, 92, 81, 87, 76, 84, 78, 79, 99, 100, 88, 77, 71, 76, 76, 83, 66, 79, 95, 85

First we need to select suitable groupings for the depth. The smallest depth at which tuna are caught is 66 m and the greatest is 95 m. So let us put each catch into a group based on a 5 m range in depth. Based on these 5 m gradations we obtain the *grouped frequency distribution* (figure 21.1).

Figure 21.1   Frequency distribution of blue-fin tuna catches

Class interval (depth in metres)	Frequency (× 100 kg)
66–70	2
71–75	9
76–80	11
81–85	7
86–90	3
91–95	3
96–100	3
	N = 38

The table now makes certain trends clearly evident. For example, most tuna are caught between depths of 71 and 80 m. The least number of tuna are caught at the shallowest depths between 66 and 70 m. You may also be asked to do some further statistical analysis of the data, for example:

Mean = 80.5 m

If we add each of the depths at which tuna are caught and divide this sum by the number of catches, this is the *average* depth at which tuna are caught.

$$75 + 67 + 76 + 73 + \ldots + 95 + 85 = 3059$$

$$3059 \div 38 = 80.5$$

Mode = 76 m

This is the depth with the highest tuna catch. The table would indicate that this occurs between 76 and 80 m, and the raw data confirms that it is at 76 m (4 catches of 100 kg).

Median = 79 m

This is the depth that has an equal number of catches above and below it. There were 18 catches below 79 m and 18 catches above 79 m, with two catches at 79 m.

75 67 76 71 73 72 77 75 75 73   79 79   86 80 80 96 93 83 81 82 92 81

73 76 78 77 71 76 76 66   87 84 99 100 88 83 95 85

Standard deviation (SD) = 8.48 m

This is the spread of the measurements from the mean. Generally, the smaller the value of the standard deviation the greater is the consistency of the data.

Drawing and using graphs

Now that you have the tuna data in the form of a table you may also wish to present the data by means of a frequency distribution diagram, which may be drawn in several ways. One of the most common is a *pie chart* (figure 21.2).

Figure 21.2   Pie chart of blue-fin tuna catches at Tuna Bay, April 2005

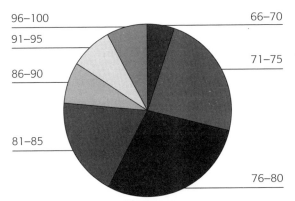

The pie chart gives a guide to the distribution of tuna catches, but it lacks detail. Another way of plotting the data is to use a *histogram* (figure 21.3).

Figure 21.3 Histogram of blue-fin tuna catches at Tuna Bay, April 2005

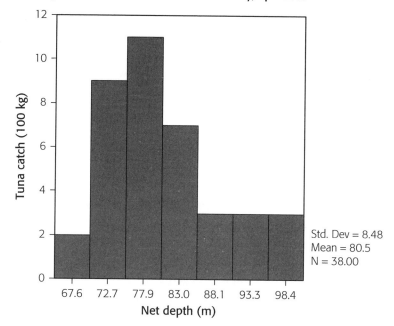

So in your report to the Tasmanian Department of Fisheries you could make recommendations to maximise the yield of tuna catches. Your report could also lend itself to some speculative interpretation. For example, why are most tuna caught within a certain depth range? Could it be that this is the depth at which a principal food source, such as sardines, is found?

This introduces an important aspect of presenting data, in the form of graphs. A graph is a *visual aid* in looking for trends, patterns, or relationships between two or more variables. Each graph should have the following:

Title

Give a succinct explanation of what the graph shows.

Relationship between the amount of added phosphate (as soluble salt) and the rate of reproduction of Escherichia coli in blood agar medium

Number

Each graph should be numbered.

Figure 2: Relationship between the amount of added phosphate (as soluble salt) and the rate of reproduction of Escherichia coli in blood agar medium

Label the axes

Both the *x* and *y* axes need to be clearly labelled showing both the numerical scale and the units. The *independent* variable, that is the one you can control or choose, is on the *x* axis. The variable that you are measuring, the *dependent* variable, is plotted on the *y* axis.

Choice of scale

The scale is the numbers on the axes and there are a variety of possibilities. A few guidelines may be useful. Choose a scale that is easy to work with and that highlights the trends. Each scale needs to cover the full range of values for both the parameters. When drawing graphs note that the scales or axes:

- need not start at zero
- need not be the same dimensions
- need to show the full range of the data

Drawing lines on graphs

When all the data points have been plotted, the next task is to decide if they can be related by some form of algebraic equation, represented as a line or curve. When plotting raw data directly onto the graph, it may be realistic to connect each data point. When the raw data has been manipulated in some way, for example the values have been averaged, then this may not be appropriate, and a different type of graph is drawn. This is most often done not by joining the data points in a point-to-point way but by looking for the *line of best fit*.

The example in figure 21.4 shows the average number of road deaths graphed against car speed.

Figure 21.4   Relationship between average road deaths in Australia and vehicle speed, 2000–04

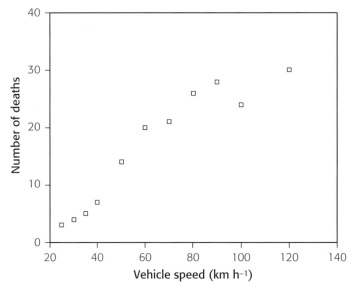

When drawing a line of best fit, a good rule is that the line drawn should have approximately an equal number of data points above and below the line, unless there is one or more obvious erroneous points. Figure 21.5 shows a possible line of best fit, in this case a linear regression analysis as calculated by the *SPSS* program. In this example, the line represents an algebraic estimate where one variable predicts the value of the second variable. The *R*-squared term is an expression of how well this estimate works, with a perfect fit being equal to one. Note that on the calculated line of best fit, there are approximately an equal number of data points above and below the line.

Figure 21.5   Relationship between average road deaths in Australia and vehicle speed, 2000–04

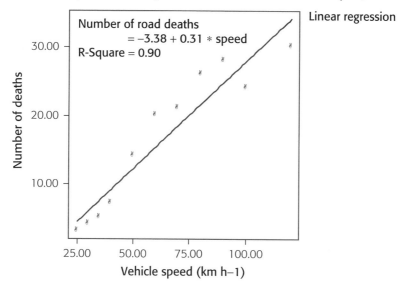

Number of road deaths
$= -3.38 + 0.31 *$ speed
R-Square $= 0.90$

Linear regression

Number of deaths

Vehicle speed (km h–1)

Drawing inferences from graphs

Let us now make some interpretations based on the data presented in the graph (figure 21.6) relating to the number of seeds produced by a flowering plant and the ground temperature during seed pod development.

Figure 21.6   Relationship between the number of seeds produced and the average ground temperature for September 2005 for the flowering plant *Eucalyptus albicans*

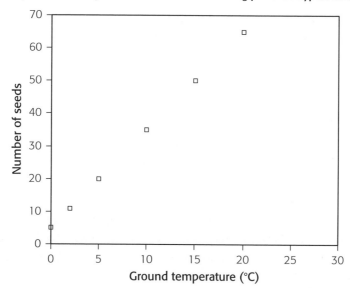

Number of seeds

Ground temperature (°C)

Is there a direct relationship between the number of seeds and the temperature? Clearly a straight line can be drawn through the data points, but what is the actual relationship? In order to determine this, we look at the rate at which Y (number of seeds) changes with respect to a change in X (ground temperature). This is referred to as the *slope* of the graph.

$$\text{Slope} = \frac{Y \text{ step}}{X \text{ step}} \quad \text{or} \quad \frac{Y_2 - Y_1}{X_2 - X_1}$$

Choosing any two points on the graph we find that

$$\text{Slope} = \frac{56 - 11}{17 - 2} = \frac{45}{15} = 3$$

Hence $Y = 3X$

The graph does not go through the origin, since when $X = 0, Y = 5$.

Therefore the true relationship between the two variables is:

$$Y = 3X + 5$$

This means that the plant produces a minimum of five seeds at a temperature of zero degrees and that for every rise in temperature of one degree there is a three-fold increase in the number of seeds produced. Further information can be found from the graph.

Interpolation

Can you find the number of seeds for a temperature not shown, but which is between two known values? For example, how many seeds are produced at 12°C? From the graph this would be about 40. You can get a more precise answer by using the equation, which would give a value of 41.

Extrapolation

What if you want to find the number of seeds for a temperature beyond the graph, for example 30°C? You can 'stretch' the graph by extending the line drawn, or use the equation.

This will give a value of 95 seeds at 30°C. In this case however, you are not absolutely sure that the same linear relationship will hold true beyond the value for which you have data, that is, 20°C. It could be that the plant reaches its maximum seed production at the temperature of 20°C, after which the seed number may decrease. Alternatively, it may increase further.

Relationships between variables

Another important and often used statistical test is the degree of a linear relationship between different variables. The name for such a relationship is the Pearson Product–Moment Correlation, but it is usually referred to as the *correlation*. The degree of the correlation is given the symbol r, *the correlation coefficient*, which may be either positive or negative, and has a maximum absolute value of one, where $r = 0$ means no correlation and $r = 1$ is a perfect correlation.

Let us consider a class of 25 first-year science students. We are interested in knowing whether or not a linear relationship exists between certain aspects of students' lives and their success at university as determined by their grade point average (GPA) (figure 21.7).

Now suppose we were interested in evaluating the relationship between students' success at university as measured by their GPA and four other factors: tertiary entrance rank (TER), hours of employment, self-rating of success, and IQ test score. Using any of the statistical packages available we could calculate the correlation between these variables (figure 21.8).

The table indicates that there is a positive correlation between GPA and each of the other variables tested, with values ranging from a low of $r = 0.115$ (hours of employment) to a high of $r = 0.516$ (TER).

Let us now look more closely at the relationship between GPA and TER. A scatter plot of these two variables (figure 21.9) indicates that as TER increases so does GPA.

From this plot we can also see that there is one point which appears incongruous with the others, and that is for the student who has a TER of 65 and a GPA of 0. Either the student has failed all topics, or withdrawn from study after the census date. To what extent is the data for this student affecting the relationship between the two variables? To test this we repeat the correlation calculations without this student's data (figure 21.10).

The table now indicates that without this student the correlation between GPA and TER has increased from $r = 0.516$ to $r = 0.686$. It has also had an impact on the correlation coefficients between the other three variables, with now a small negative correlation between GPA and hours of employment. The question now is whether to exclude the data for student as an anomaly or whether to include it.

Warning

A final word when interpreting the meaning of a correlation coefficient. A correlation between two variables, A and B, is *not* an indication of cause and effect. Though the two variables may show a correlation, variable A does not cause variable B, or vice versa. Looking back to the shoe size and death rate correlation mentioned in the introduction, we can see that, though these two factors are related, having large feet does not cause soldiers to die in battle.

Figure 21.7   Data for students in a first-year science class in 2006 (N = 25)

ID	TER	GPA	Sex	Age	School	Work	Success	IQ test
300001	70.00	5.25	F	21	1	0	7	105
300002	56.75	4.25	M	29	1	0	8	80
300003	82.00	5.75	F	18	1	1	7	95
300004	94.05	6.25	F	20	2	1	9	110
300005	72.00	6.00	F	27	1	2	7	95
300006	87.00	6.25	M	19	3	0	8	105
300007	58.00	5.33	M	38	2	1	7	94
300008	71.10	5.33	F	19	1	2	6	96
300009	92.00	6.00	F	46	1	4	8	99
300010	89.50	5.00	F	19	2	2	7	96
300011	97.65	7.00	M	18	3	1	9	121
300012	77.40	5.75	F	19	1	0	7	109
300013	98.70	5.00	F	19	1	0	9	120
300014	67.28	4.67	F	18	2	1	7	108
300015	73.03	5.00	M	19	1	0	8	100
300016	92.70	6.00	F	19	3	1	8	119
300017	70.00	4.75	F	18	1	2	6	106
300018	86.60	5.00	F	18	2	2	7	105
300019	88.00	5.00	F	19	1	1	8	104
300020	69.85	2.66	F	19	2	0	7	101
300021	65.00	.00	F	34	3	0	6	98
300022	77.40	4.00	F	46	1	3	7	99
300023	65.00	6.00	F	26	1	1	7	96
300024	44.40	3.00	M	23	1	2	6	89
300025	76.90	4.50	F	22	1	4	8	94

Code

ID = student number (a fictitious number is used)

TER = tertiary entrance rank

GPA = grade point average for first-year studies (0 = lowest, to 7 = highest)

Sex = sex of student (1 = male, 2 = female)

Age = age in years on January 1 of first-year at university

School = school attended (1 = public high school, 2 = private school, 3 = other school)

Work = number of hours per/week working (0 = 0 hours, 1 = 1–5 hours, 2 = 6–10 hours, 3 = 11–15 hours, 4 = 16–20 hours, 5 = more than 20 hours)

Success = student's own predicted success at university (1 = lowest to 10 = highest)

IQ test = score on IQ test

Figure 21.8   Correlation coefficients (N = 25)

	GPA	TER	Work	Success	IQ test
GPA	1				
TER	0.516	1			
Work	0.115	0.090	1		
Success	0.511	0.679	−0.121	1	
IQ test	0.358	0.703	−0.234	0.471	1

Figure 21.9   Relationship between TER and GPA for first-year science students

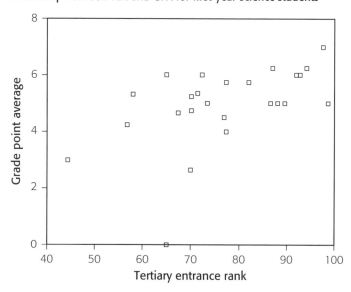

Figure 21.10   Correlation coefficients between variables for first-year science students (N = 24)

	GPA	TER	Work	Success	IQ test
GPA	1				
TER	0.686	1			
Work	−0.060	0.054	1		
Success	0.434	0.667	−0.203	1	
IQ test	0.435	0.702	−0.259	0.471	1

Statistical terminology

Figure 21.11 lists some of the common statistical terminology which you may come across in your studies.

Figure 21.11   Common statistical terminologies and their meaning

Terminology	Meaning
Hypothesis testing	an inferential procedure that uses data to evaluate the credibility of an hypothesis about a given population
Null hypothesis, H_0	an hypothesis that states that an independent variable has no effect on the dependent variable
Alternative hypothesis, H_1	an hypothesis that states that an independent variable has an effect on the dependent variable
Type I error	rejection of a null hypothesis when it is true
Type II error	failure to reject a null hypothesis when it is false
Variance	the mean of the sum of the squared amounts by which values deviate from the mean value
Normal distribution	a bell-shaped curve that represents the distribution of the values of a variable about the mean
One-sample T test	the procedure that tests whether the mean of a single variable differs from a specified constant
The independent-samples T test	a procedure that compares means for a single variable between two independent sample groups
The paired-samples T test	a procedure that compares the means of two variables for a single group: it shows the differences between two variables for each case and tests whether the average differs from 0
ANOVA (analysis of variance)	a procedure that produces a one-way analysis for a quantitative dependent variable by a single independent variable: analysis of variance is used to test the hypothesis that several means are equal
Regression	a technique for finding the best line of fit for the relationship between two variables such that $y = ax + b$
Chi-square (X^2) test	a procedure to test a hypothesis about how well a sample population fits the population proportions specified by a null hypothesis

Summary

This chapter has been a very brief introduction to some of the statistical techniques you are likely to come across in your studies. Even the most descriptive areas of science use numbers to describe patterns or relationships, so having an understanding of some of these fundamentals can be of great assistance to you in your studies.

Appendixes

1. Key Words in Written Assignment Questions

The following list gives a brief explanation of some of the key terms you may come across in written assignment questions.

Analyse	Examine critically various components and determine how they are related in order to gain understanding
Argument	A presentation of your reasons for or against a proposition using evidence to support your case
Compare	Point out similarities or differences
Conclude	Draw together different parts and make a final assessment
Contrast	Stress the differences in characteristics or qualities
Criticise	Provide a judgment as to the strengths or weaknesses or an argument or hypothesis, using evidence to support your view
Define	Explain the meaning or essential qualities
Demonstrate	Illustrate using specific examples
Describe	State the characteristics, features, or outline of some phenomenon with minimal interpretation
Discuss	Examine critically and in depth an issue using argument supported by evidence
Evaluate	Appraise value or worth giving both advantages and disadvantages
Examine	Investigate a concept and critically analyse its value
Explain	Clarify the meaning of a particular phenomenon
Extrapolate	Infer what is not known from given data and information, or estimate the value beyond known limits
Generalise	Make a comprehensive statement about data or information
Illustrate	Use an example as a means of explaining a phenomenon
Interpret	Give meaning using your judgment and known information or values

Justify   Prove by giving your reasons for a particular conclusion, decision, or perspective

Outline   Organise information into a cohesive arrangement or classification

Summarise   Give the main points or facts in a condensed form excluding specific details

Synthesise   Take evidence from different sources to develop a conclusion or perspective

2. Unusual Plurals

In most cases to make a word plural (to mean more than one), add an 's' or 'es'; however, there are many exceptions. Some that you may find in science topics include:

Singular	Plural
alga	algae
analysis	analyses
antenna	antennae
appendix	appendices/appendixes
bacterium	bacteria
criterion	criteria
datum	data
formula	formulae
genus	genera
hypothesis	hypotheses
locus	loci
medium	media
mouse (computer)	mouses
nucleus	nuclei
ovum	ova
phenomenon	phenomena
quantum	quanta
radius	radii
schema	schemata
species	species
spectrum	spectra
stimulus	stimuli
vortex	vortices

When referring to periods of time, write these as, for example, the 1900s, not the 1900's. Plurals of abbreviations add an 's' to the abbreviation. For example, CDs is the plural of CD (compact disc), and PCBs is the plural of PCB (polychlorinated biphenyl).

3. Prefixes

Many terms in science use a Latin or Greek prefix as an integral part of the word. The following are some of the most commonly used prefixes.

Prefix	Meaning	Example
a, an	without	anorexic, without appetite
ante	before	antenatal, before birth
anti	against	antidote, neutralises a toxin
auto	self	autoimmume, self immune
bi	two	bipedal, with two legs
contra	against	contradictory, an opposing view
dys	ill, poor	dysfunctional, functioning poorly
e, ex, exo	out (of)	expiration, to breathe out
endo	within	endocrine, glands that secret within
extra	beyond	extracellular, outside the cell
hemi	half	hemispherical, half the sphere
hyper	above	hypertension, pressure (tension) above normal
hypo	below	hypotension, pressure (tension) below normal
inter	between	international, between nations
intra	within	intracellular, within the cell
iso	same	isotonic, of the same strength
mono	one, single	mononuclear, with one nucleus
multi	many	multicellular, of many cells
omni	all	omnipotent, all powerful (potent)
para	aside, beyond	paranormal, outside the normal
poly	many	polyethylene, a plastic of many ethylene units
post	after	postnatal, after birth
pre	before	premenstrual, before menstruation
pseudo	false	pseudomembranous, a false membrane
semi	half	semiconductor, a partial conductor
syn	together	synthesise, to make together
trans	across	translocate, to move across
ultra	beyond	ultraviolet radiation, beyond the wavelength of violet light

4. The Use of Apostrophes

The following is based on *The Oxford Guide to Style*, 2nd edn, 2002.

Possession

An apostrophe is most frequently used to indicate singular or plural possessive. When using possessive nouns there are a number of basic principles:

1   If a singular noun does not end in s:

    •   The delivery *man's* truck was a Kenworth.

    •   Mr *Howard's* speech on education was limp and fatuous.

    •   The *student's* attempt to solve the problem was rewarded.

2   There seems to be no universally accepted format when a singular common noun ends in s, but in the majority of cases add **'s**

    •   *Mars's* gravitational force is less than that on Earth.

    •   The *witness's* story did not match the events recorded on tape.

3   If a noun is plural in form and ends in an s, add an apostrophe only, even if the intended meaning of the word is singular (such as mathematics and measles.)

    •   The instructor asked us to analyse the *problems'* meanings.

    •   The tutor had us check all of the *rats'* cages.

4   If a plural noun does not end in s:

    •   Many social activists are concerned with *children's* rights.

    •   The *media's* coverage of the cricket is very good.

5   If there is joint possession, use the correct possessive for only the possessive closest to the noun.

    •   Abbott and *Costello's* article was difficult to understand.

    •   George was worried about his mother and *father's* health.

6   If there is a separate possession of the same noun, use the correct possessive form for each word.

    •   The *owner's* and the *boss's* cars are the same make.

    •   The *rats'* and the *possums'* cages all needed cleaning.

7 In a compound construction, use the correct possessive form for the word closest to the noun. Avoid possessives with compound plurals.

- My *father-in-law's* Ferarri is red.
- The *forest ranger's* truck is an old Toyota.

8 Personal names.

There is considerable confusion as to the use of an apostrophe with personal names depending on whether the name ends with an s or not. For consistency, always add **'s**.

- *Chris's* exam scores were higher than any other student.
- *Burns's* results are very accurate.

9 When referring to inanimate objects there is often no need for an apostrophe:

- The door of the car = the car door.
- The leg of the table = the table leg.

10 Many non-English words that end with a silent 's' or 'x' so the preferred style is:

- Alexander Dumas's first novel was well received.
- The Bordeaux's bouquet is legendary.

11 When referring to expressions of time, the convention is to add **'s** for singular time and **'** for plurals:

- It was all in a day's work.
- Tomorrow I am going on a fortnight's holiday.
- Can you get me a copy of today's newspaper?
- I am owed three weeks' wages.

Omission of letters

An apostrophe is also used in a contracted word to indicate a missing letter or letters. However, in academic writing these contractions should be avoided.

Contracted form	Preferred form
It's a well-known fact	It is a well-known fact
Don't run in the labs	Do not run in the labs
I'm not attending the lecture	I am not attending the lecture
He'll do the experiment later	He will do the experiment later
Who's next to talk	Who is next to talk
We shouldn't see any change	We should not see any change
The experiment didn't work	The experiment did not work
We could've done that tutorial	We could have done that tutorial

5. SI Units and Fundamental Constants

The Greek Alphabet

Greek name	Greek name	English equivalent	Greek letter	Greek name	English equivalent
A α	Alpha	a	N υ	Nu	n
B β	Beta	b	Ξ ξ	Xi	x
Γ γ	Gamma	g	O o	Omicron	o
Δ δ	Delta	d	Π π	Pi	p
E ε	Epsilon	e	P ρ	Rho	r
Z ζ	Zeta	z	Σ σ	Sigma	s
E ε	Eta	e	T τ	Tau	t
Θ θ	Theta	th	Υ υ	Upsilon	u
I ι	Iota	i	Φ φ	Phi	ph
K κ	Kappa	k	X χ	Chi	ch
Λ λ	Lambda	l	Ψ ψ	Psi	ps
M μ	Mu	m	Ω ω	Omega	o

Numerical prefixes

Number	Greek	Latin
0.5	hemi	semi
1.0	mono	uni
1.5	–	sesqui
2	di	bi
3	tri	ter
4	tetra	quadric
5	penta	quinque
6	hexa	sexi
7	hepta	septi
8	octa	octo
9	ennea	nona
10	deca	deci
Many	poly	multi

Decimal fractions and multiples

Number	Prefix	Symbol	Number	Prefix	Symbol
10^{-18}	atto	a	10^{1}	deca	da
10^{-15}	femto	f	10^{2}	hecto	h
10^{-12}	pico	p	10^{3}	kilo	k
10^{-9}	nano	n	10^{6}	mega	M
10^{-6}	micro	μ	10^{9}	giga	G
10^{-3}	milli	m	10^{12}	tera	T
10^{-2}	centi	c	10^{15}	peta	P
10^{-1}	deci	d	10^{18}	exa	E

The SI Units

| Physical quantity | | Units | |
Name	Symbol	Name	Symbol
Length	l	Metre	m
Mass	m	Kilogram	kg
Time	t	Second	s
Electrical current	I	Ampere	A
Temperature	T	Kelvin	K
Luminous intensity	lv	Candela	cd
Amount of substance	n	Mole	mol

Common derived units

Physical Quantity		Units		
Name	Symbol	Name	Symbol	Definition
Area	A	square metres		m^2
Volume	V	cubic metres		m^3
Velocity	v	metres per second		$m\ s^{-1}$
Density	ρ	mass per unit volume		$kg\ m^{-3}$ or $g\ cm^{-3}$
Acceleration	a	rate of change of velocity		$m\ s^{-2}$
Frequency	υ	oscillations per second	Hz	s^{-1}
Energy	E	joules	J	$kg\ m^2\ s^{-2}$
Force	F	newton	N	$kg\ m\ s^{-2} = J\ m^{-1}$
Pressure	p	pascal	Pa	$kg\ m^{-1}\ s^{-2} = N\ m^{-2}$
Power	P	watt	W	$kg\ m^{-1}\ s^{-3} = J\ s^{-1}$
Electric charge	Q	coulomb	C	$A\ s$
Electric potential difference	U	volt	V	$kg\ m^2\ s^{-3}\ A^{-1} = J\ A^{-1}s^{-1}$
Electric resistance	R	ohm	Ω	$kg\ m^2\ s^{-3}\ A^{-2} = V\ A^{-1}$
Electric conductance	G	siemens	S	$kg^{-1}\ m^{-2}\ s^3\ A^2 = \Omega^{-1}$
Electric capacitance	C	farad	F	$A^2\ s^4\ kg^{-1}\ m^{-2} = A\ s\ V^{-1}$
Magnetic flux	ϕ	weber	Wb	$kg\ m^2\ s^{-2}\ A^{-2} = V\ s$
Inductance	L	henry	H	$kg\ m^2\ s^{-2}\ A^{-2} = V\ A^{-1}\ s$
Magnetic flux density	B	Tesla	T	$kg\ s^{-2}\ A^{-1} = V\ s\ m^{-2}$

Fundamental constants

Description	Symbol	Value
Avogadro's constant	N_A	6.022169×10^{23} mol^{-1}
Faraday's constant	F	96486 C mol^{-1}
Charge of electron	e	$1.6021917 \times 10^{-19}$ C
Mass of electron	m_e	9.109558×10^{-31} kg
Mass of proton	m_p	1.672614×10^{-27} kg
Mass of neutron	m_n	1.674920×10^{-27} kg
Planck's constant	h	6.626196×10^{-34} J s
Speed of light in vacuum	c_0	2.9979250×10^{8} m s^{-1}
Permeability of vacuum	μ_o	$4\pi \times 10^{-7}$ m s^{-2} A^{-2}
Permittivity of vacuum	ε_o	$8.8541853 \times 10^{-12}$ kg^{-1} m^{-3} s^4 A^2
Bohr radius	a_0	$5.2917715 \times 10^{-11}$ m
Rydberg constant	R_H	1.09677578×10^{7} m^{-1}
Bohr magneton	μ_B	9.274096×10^{-24} A m^2
Gas constant	R	8.31434 J K^{-1} mol^{-1} 1.98717 cal K^{-1} mol^{-1} 82.0562 cm^3 atm K^{-1} mol^{-1}
Boltzmann's constant	k	1.380622×10^{-23} J K^{-1}
Volume of ideal gas	V_m	22413.6 cm^3 mol^{-1} at 1 atm, 0° C 24465.1 cm^3 mol^{-1} at 1 atm, 25° C
Base of natural logarithm	e ln x π	2.71828 ... 2.3026 log x 3.14159265 ...

Some common functions used in calculus

Differentiation		Integration	
function/relationship	**math-speak**	**function/relationship**	**math-speak**
distance, $s(t) \rightarrow$ speed, $v(t)$	$v(t) = \frac{ds}{dt}$	speed, $v(t) \rightarrow$ distance, $s(t)$	$s = \int v(t)dt$
speed, $v(t) \rightarrow$ acceleration, $a(t)$	$a(t) = \frac{dv}{dt}$	acceleration, $a(t) \rightarrow$ speed, $v(t)$	$v = \int a(t)dt$
mass, $m(x) \rightarrow$ density, $\rho(x)$	$\rho(x) = \frac{dm}{dx}$	density, $\rho(x) \rightarrow$ mass, $m(x)$	$m = \int \rho(x)dx$
population, $P(t) \rightarrow$ growth, $r(t)$	$r(t) = \frac{dP}{dt}$	growth, $r(t) \rightarrow$ population, $P(t)$	$r = \int \frac{dP}{dt} dt$

Bibliography

Listed here are those books we have consulted and others which we believe are worth your consideration. Many of these books should be available through university libraries or through the university bookstore on your campus. Amazon.com is also a good source of academic books.

Academic skills for university

Biber, D. 2006, *University language*, John Benjamins, Amsterdam.

Bullen, M. & Janes, D.P. 2006, *Making the transition to e-learning*, Information Science Publishing, Hershey PA.

Cooper, G. 2003, *The intelligent student's guide to learning at university*, Common Ground, Altona.

Covey, S. 1989, *The seven habits of highly effective people*, The Business Library, Melbourne.

Fairbairn, G.J. 1996, *Reading, writing and reasoning*, 2nd edn, Open University Press, Buckingham UK.

Fry, R. 2004, *How to study*, 6th edn, Thomson Delmar Learning, Clifton Park, NY.

Game, A., & Metcalfe, A. 2003, *The first-year experience*, The Federation Press, Sydney.

Hay, I., Bochner, D., & Dungey, C. 2006, *Making the grade: a guide to successful communication and study*, 3rd edn, Oxford University Press, Melbourne.

Marshall, L.A. 1998, *A guide to learning independently*, Open University Press, Buckingham, UK.

Wallace, A., Schirato, T., & Bright, P. 1999, *Beginning university: thinking, researching and writing success*, Allen & Unwin, Crows Nest, NSW.

Williams, L., & Germov, J. 2001, *Surviving first-year uni.*, Allen & Unwin, Crows Nest, NSW.

Critical thinking and argument

Allen, M. 1997, *Smart thinking: skills for critical understanding and writing*, Oxford University Press, Oxford.

Audi, R. (ed.)1999, *The Cambridge dictionary of philosophy*, 2nd edn, Cambridge University Press, Cambridge UK.

Couvalis, G. 1997, *The Philosophy of science: science and objectivity*, Sage Publications, London.

Facione, P. 1998, *Critical thinking: what it is and why it counts*, California Academic Press.

Halpern, D.H. 1997, *Critical thinking across the curriculum*, Lawrence Erlbaum Associates, New Jersey.

Nickerson, R.S. 1986, *Reflections on reasoning*, Lawrence Erlbaum Associates, New Jersey.

Paul, R., & Elder, L. 2001, *The miniature guide to critical thinking, concepts, and tools*, The Foundation for Critical Thinking, Dillon Beach, CA.

van den Brink-Budgen, R. 2000, *Critical thinking for students: learn the skills of critical assessment and effective argument*, 3rd edn, How To Books, Oxford.

Research and communication skills

Alley, M. 2005, *The craft of scientific presentations: critical steps to succeed and critical errors to avoid*, Springer, New York.

Berkman, R. 1994, *Find it fast: how to uncover expert information*, Harper Perennial, New York.

Booth, V. 1993, *Communicating in science: writing a scientific paper and speaking at scientific meetings*, 2nd edn, Cambridge University Press, Cambridge.

Bouma, G.D. 2000, *The research process*, 4th edn, Oxford University Press, Oxford.

Gunther, N. 1982, *The art of effective speaking*, Reed, Sydney.

Higgs, J., Sefton, A., Street, A., McAllister, L., & Hay, I. 2005, *Communicating in the health and social sciences*, Oxford University Press, Melbourne.

Hiltz, S.R., & Goldman, R. (eds) 2004, *Learning together online: research on asynchronous learning networks*, Lawrence Erlbaum Associates, Mahwah, New Jersey.

Klein, R., Hunt, M., & Lee, R. 1999, *The essential workbook for library and Internet research*, McGraw-Hill, New York.

Mann, C., & Stewart, F. 2000, *Internet communication and qualitative research: a handbook for researching online*, Sage Publications, London.

Mohan, T., McGregor, H., Saunders, S., & Archee, R. 2004, *Communicating as professionals*, Thomson, Sydney.

Schlein, A.M. 2004, *Find it online: the complete guide to online research*, 4th edn, Facts on Demand Press, Tempe AZ.

Seeley, J. 2000, *The Oxford guide to writing & speaking*, Oxford University Press, Oxford.

Shortland, M., & Gregory, J. 1991, *Communicating science: a handbook*, Longman Scientific & Technical, New York.

Learning more about learning

Anderson, J.R. 1999, *Cognitive psychology and its implications*, WH Freeman & Company, New York.

Anderson, L.W., Krathwohl, D.R., Airasian, P.W., Cruikshank, K.A., Mayer, R.E., Pintrich, P.R., Raths, J., & Wittrock, M.C. 2001, *Taxonomy for learning, teaching, and assessing: a revision of Bloom's taxonomy of educational objectives*, Abridged Edition, Allyn & Bacon, New York.

Biggs, J.B. 1987, *Student approaches to studying and learning*, Australian Council for Educational Research, Hawthorn, Australia.

Biggs, J.B. 1999, *Teaching for quality learning at university*, SRHE and Open University Press, Buckingham.

Dunn, R., & Griggs, S.A. 2000, *Practical approaches to using learning styles in higher education*, Bergin Garvey/Greenwood, Westport, CT.

Gibbs, G. (ed.) 1994, *Improving student learning: theory and practice*, Oxford Centre for Staff Development, Oxford.

Kolb, D.A. 1984, *Experiential learning: experience as the source of learning development*, Prentice-Hall, Engelwood Cliffs NJ.

Prosser, M., & Trigwell, K. 1999, *Understanding learning and teaching*, SRHE and Open University Press, Buckingham.

Richardson, J.T.E. 2000, *Researching student learning*, SRHE and Open University Press, Buckingham.

Quantitative methods: mathematics, statistics, and problem solving

Bland, M. 2000, *An introduction to medical statistics*, 3rd edn, Oxford University Press, Oxford.

Courant, R., Robbins, H., & Stewart, I. 1996, *What is mathematics? An elementary approach to ideas and methods*, 2nd edn, Oxford University Press, New York.

Dombroski, T.W. 2000, *Creative problem solving: the door to individual success and change*, ExcelPress, Lincoln NE.

Everitt, B.S., & Hothorn, T. 2006, *A handbook of statistical analysis*, Taylor and Francis, Boca Raton Fl.

Eves, H. 1997, *Foundations and fundamental concepts of mathematics*, 3rd edn, Dover Publications, Mineola NY.

Fobes, R. 2006, *Creative problem solver's toolbox*, Creative Solutions, Portland OR.

Greer, B., & Mulhern, G. 2001, *Making sense of data and statistics in psychology*, Palgrave Macmillan, New York.

Nation, D.S., & Siderman, S.J. 2004, *Mathematics problem solving coach*, EVP Publishing, New York.

Scott, K.S. 1995, *Beginning mathematics for chemistry*, Oxford University Press, Oxford.

Steiner, E. 1996, *The chemistry maths book*, Oxford University Press, Oxford.

Referencing styles

American Psychological Association 2001, *Publication manual of the American Psychological Association*, 5th edn, Washington.

Commonwealth of Australia 2002, *Style manual for authors, editors and printers*, 6th edn, John Wiley & Sons Australia Ltd.

Council of Biological Editors, CBE Style Committee 1983, *CBE style manual: a guide for authors, editors and publishers in the biological sciences*, 5th edn, Bethesda, Md.

Ritter, R.M. 2002, *The Oxford guide to style*, Oxford University Press, Oxford.

Turabian, K., Grossman, J., & Bennett, A. 1996, *The Chicago manual of style: a manual for writers of term papers, theses and dissertations*, 16th edn, University of Chicago Press, Chicago.

Writing for the sciences: essays, assignments, reports

Alley, M. 1996, *The craft of scientific writing*, 3rd edn, Springer-Verlag Inc., New York.

Anderson, J., & Poole, M. 2000, *Thesis and assignment writing*, 3rd edn, Wiley, Queensland.

Barrass, R. 2002, *Scientists must write: a guide to better writing for scientists, engineers, and students*, Routledge, London.

Burton, L.J. 2002, *An interactive approach to writing essays and research reports in psychology*, John Wiley & Sons, Milton, Queensland.

Day, R.A. 2005, *How to write and publish a scientific paper*, 5th edn, Oryx Press, Phoenix, AZ.

Ebel, H.F., Bliefert, C., & Russey, W.E. 1987, *The art of scientific writing*, VCH Publishers, Weinheim, FRG.

Lindsay, D. 1995, *A guide to scientific writing*, 2nd edn, Longman Cheshire, Melbourne.

Matthews, J.R., Bowen, J.M., & Matthews, R.W. 2001, *Successful scientific writing: a step-by-step guide for biomedical scientists*, 2nd edn, Cambridge University Press, Cambridge.

Moore, R. 1995, *Writing to learn science*, Brooks-Cole Publishing, Fort Worth.

Plotnik, A. 1984, *The elements of editing*, Macmillan Publishing Company, New York.

Strunk, W. Jr, & White, E.B. 2000, *The elements of style*, Allyn and Bacon, USA.

Trelease, S.F. 1969, *How to write scientific and technical papers*, MIT Press, Cambridge, Ma.

Truss, L. 2004, *Eats, shoots & leaves*, Profile Books, London.

Van Alstyne, J. 2001, *Professional and technical writing strategies*, 5th edn, Prentice Hall Inc., New Jersey.

Winckel, A., & Hart, B. 2002, *Report writing style guide for engineering students*, 4th edn, University of South Australia, Adelaide.

Online Resources

The rapid growth of the Internet has resulted in many thousands of websites with information relevant to tertiary study. Some of these are very good and many are not. The sites listed here are mainly from universities in Australia, Britain, and North America, and are those we believe to be most relevant to the skills we have attempted to outline in this book. The Web addresses were correct at the time of writing but no assurances can be given that they have not since changed.

Academic and study skills

Most universities have a dedicated site on academic skills as well as links to other useful websites. The following are just a few worth investigating:

http://www.flinders.edu.au/slc

http://www.usq.edu.au/ltsu/alsonline/default.htm

http://startup.curtin.edu.au/study/index.cfm

http://www.ucalgary.ca/~dmjacobs/study_skills_sites.html

http://www.sussex.ac.uk/languages/1-6-8.html

http://www.jcu.edu.au/studying/services/studyskills/online

http://www.studyskills.soton.ac.uk/

http://people.brunel.ac.uk/%7emastmmg/ssguide/sshome.htm

Critical thinking and argument

An interactive site for argumentation and critical thinking from Humboldt University

http://www.humboldt.edu/~act/HTML/

The essentials of critical thinking from Canberra University

http://www.canberra.edu.au/studyskills/learning/crithink.html#what

The Austhink critical thinking homepage

http://www.austhink.org/critical/

Critical Thinking Links 2006

http://www.nvcc.edu/home/jtrabandt/discussion/criticalthinking/ctlinks.html

Critical Thinking Skills—Definitions and Assessment

http://ericae.net/faqs/crit_tnk.htm.

Critical thinking across disciplines

http://www.nvcc.edu/home/nvnaqud/critical_thinking/

Deductive and inductive reasoning

http://www.socialresearchmethods.net/kb/dedind.htm

E-learning

There are literally hundreds of websites that can assist you in using the Web and other electronic resources. We have listed here only those we believe to be particularly good.

This site at Berkeley University is a good general introduction to using search engines

http://www.lib.berkeley.edu/TeachingLib/Guides/Internet/FindInfo.html

The following site is set up as an online tutorial which takes you step-by-step through the different aspects of Web searching

http://www.thelearningsite.net/cyberlibrarian/searching/ismain.html

A very basic but user-friendly guide to using the Internet

http://www.learnthenet.com/english/index.html

A guide to net etiquette

http://www.albion.com/netiquette/corerules.html

English language and editing

Dictionaries and thesaurus

http://www.facstaff.bucknell.edu/rbeard/diction.html

http://www.thefreedictionary.com

http://search.thesaurus.com/

http://www.hti.umich.edu/dict/oed/

http://humanities.uchicago.edu/forms_unrest/ROGET.html

Editing

http://www.jcu.edu.au/studying/services/studyskills/editing.pdf

http://owl.english.purdue.edu/handouts/general/gl_edit.html

Grammar

http://webster.commnet.edu/grammar/index.htm

http://grammar.ccc.commnet.edu/grammar/

http://www.grammarbook.com

http://www.eslcafe.com/quiz/

Punctuation

http://www.sussex.ac.uk/langc/skills/punc.html

Intelligence tests

There are many electronic sites that purport to present intelligence or IQ tests. Most of these are merely commercial sites wishing to separate you from your money. The following are just three of the better sites with online tests:

http://www.intelligencetest.com/

http://www.2h.com/iq-tests.shtml

http://www.queendom.com/

Learning styles and approaches

There are several electronic sites that describe some of the more common learning styles. Some of these allow you to do online evaluations.

This site gives a very informative overview of many different aspects of teaching and learning

http://www.learningandteaching.info/learning/contents.htm

A version of the Felder and Silverman Index of Learning Styles giving immediate feedback

http://www.crc4mse.org/ILS/Index.html

An overview of the Honey and Mumford Learning Style Preferences instrument

http://www.campaign-for-learning.org.uk/aboutyourlearning/whatlearning.htm

A description of the Index of Learning Styles instrument used to assess learning preferences

http://www.ncsu.edu/felder-public/ILSpage.html

The Mindmedia test for learning preferences allows you to do the test online

http://www.mindmedia.com/brainworks/profiler

A comprehensive overview of the Learning and Thinking Styles Inventory

http://admin.vmi.edu/ir/ltsi.htm#Sensory%20Preference

Mathematics and statistics

The number of websites available which have good interactive information on the use of mathematics or statistics is fairly limited. Listed are some of the better examples.

The Mathcentre in the UK has a guide to maths skills for various sciences

http://www.mathscentre.ac.uk/students.php

An introduction to statistical procedures developed by David Stockburger at Southwest Missouri State University

http://www.fmi.uni-sofia.bg/fmi/statist/education/IntroBook/sbk13toc.htm

The home page of Statsoft

http://www.statsoft.com/textbook/stathome.html

Hyperstat online home page, a resource developed by David Lane

http://davidmlane.com/hyperstat/

Presentation skills: talks and posters

There are several sites dedicated to presentations skills; the following are good examples.

Oral presentations

http://www.cs.wisc.edu/~markhill/conference-talk.html#badtalk

http://web.cba.neu.edu/~ewertheim/skills/oral.htm#strategy

http://www.surrey.ac.uk/Skills/pack/pres.html

Poster presentations

http://people.eku.edu/ritchisong/posterpres.html

http://www.ncsu.edu/project/posters/IndexStart.html

http://www.canberra.edu.au/studyskills/learning/oral

Referencing styles

Your own university is likely to have Web pages dedicated to the different forms of referencing systems used within your university. It is a good idea to look at these first. In case you need additional examples, we suggest the following:

APA style

www.library.uq.edu.au/training/citation/apa.pdf

http://www.apastyle.org

Author–date (Harvard) style

http://www.lib.flinders.edu.au/resources/sub/healthsci/referencing/

CBE style

http://www.bedfordstmartins.com/online/cite8.html

Chicago style

http://www.wisc.edu/writing/Handbook/DocChicago.html

Oxford style (Footnotes)

http://www.usq.edu.au/library/help/ehelp/ref_guides/oxford.htm

Vancouver style

http://www.lib.monash.edu.au/tutorials/citing/vancouver.html

Scientific and technical writing

There are many websites dedicated to writing skills, some are very general, while others are designed with particular disciplines in mind. The following should be of value to you.

A very comprehensive writing and presentation site with examples of just about everything

http://www.writing.eng.vt.edu/index.html

One of the best and most extensive guides to English language was developed by the late Charles Darling of Capital Community College

http://grammar.ccc.commnet.edu/grammar/

General writing skills

http://www.canberra.edu.au/studyskills/writing

http://www.sussex.ac.uk/languages/1-6-8-7.html

http://www.flinders.edu.au/SLC/Brochures/writing_an_essay.pdf

http://owl.english.purdue.edu/handouts/

http://www.indiana.edu/~cheminfo/ca_swa.html

Report writing

http://www.unisanet.unisa.edu.au/learningconnection/student/learningAdvisors/
          documents/report-writing-engineering.pdf

http://www.sussex.ac.uk/engineering/1-3-11-2.html

http://encs.concordia.ca/scs/Forms/Form&Style.pdf

Scientific and technical writing

http://www.writing.eng.vt.edu/exercises/

http://www.rbs0.com/tw.htm

http://www.io.com/~hcexres/textbook/

Time management

Every university seems to have a site dedicated to aspects of time management, indicating the importance of this skill to all students.

Many different aspects of time management strategies from Dartmouth University

http://www.dartmouth.edu/~acskills/success/time.html

An online exercise in your time management skills from Ohio State University

http://studytips.aac.ohiou.edu/?Function=TimeMgt&Type=168hour

Twenty steps to successful time management from Cornell University

http://www.clt.cornell.edu/campus/learn/LSC%20Resources/20stepstotimemgmt.pdf

How to manage time and to set priorities

http://marin.cc.ca.us/%7Edon/Study/5time.html

Index